EDWARD AGO

GIANT EUSOCIAL INSECTOIDS AND THEIR ANTHROP SYMBIONTS

ISBN **978-1-4710-5668-0**

The author's spelling and punctuation
have been preserved in the text.

Published in Canada
by Altaspera Publishing & Literary Agency Inc.

About topic. This book by biologist Edward Agon, PhD in Biochemistry, «Giant Eusocial Insectoids and their Anthrop Symbionts» is devoted to biological and some social aspects of anomalistics, source of strange phenomena, nature of Aliens and their real place in the terrestrial biosphere.

THANKS

Author's thanks to Mr. Mironov, Nikolai Subbotin, Anton Anfalov, Mikhail Gershtein, Sergey Malcev, Oleg Obiedkov, Sergiy Shevik, Elen Alkor, Vladimir Litovka, Valentin Konon, and Alexei Burmistrov for help, advices and materials.

CONTENT

FOREWORD TO ENGLISH EDITION

We're in a hurry to disappoint dreamers of interstellar contact. What we know about UFOs is just «tip of iceberg». On Earth, besides us, there is a powerful non-human civilization. Abnormal phenomena are caused by terrestrial creatures belonging to an intelligent ancient species. This species is much older than human. Its main representatives are not at all «extraterrestrial humanoids», as is commonly believed. Formative species of this alien civilization are termitoids (insectoids). This is an ancient non-human species, which has a completely terrestrial evolutionary origin. Due to its natural abilities, antiquity and high technology, as well as symbiosis with genetically modified Anthrops, this species dominates not only on Earth, but also, possibly, on other planets.

There is reason to believe that appearance of different castes of termitoids has been observed for a long time and that they are still encountered today. Sometimes they are mistaken for «reptilians» or insectoids. These include such well-known phenomena as «chupacabra», Jersey Devil, Mothman, and less well-known ones, such as Flatwood monster, Jumping Jack (Spring-heeled Jack), and others. Quite often, witnesses observed «humanoids» of various sizes and types.

This book is dedicated to the study of the nature of the species, which is erroneously considered to be «extraterrestrial». We, humans, have coexisted with them, without knowing anything, for many thousands of years. We suffer from their raids on our farms, from mutilations of domestic animals and from abductions of people. This does not exhaust his influence on Humanity. Some people saw and described something like huge insects. Perhaps in order to disguise this fact, which is well known to secret services of different countries, films and cartoons with alien characters, including insectoids, are being created.

The author, PhD in biology, attempted to acquaint the reader with those who actually live with us on Earth, or rather underground of the Earth. Below we present an analysis of one document that accidentally got on the Internet after the collapse of the USSR and its all-powerful special services. This document is hardly understand, especially for those who do not have a special biological or medical education. Indeed, until now, those who have it, do not believe in existence of Aliens, are afraid to ruin their careers or be branded a fool.

BASIC CONCEPTS

ANOMALISTICS – the study of phenomena unknown to science. UFOLOGY – a section of anomalistics that studies UFOs and Alien civilizations. UFO (unidentified flying object) – we mean a vehicle of Aliens and Anthropes.

ALIENS are intelligent Giant Eusocial Insectoids whose populations include several castes: Higher, Workers, Soldiers.

ANTHROPES – anthropoids (humanoids, EBE), anthropomorphic creatures, symbionts of Aliens.

ALIEN ENVIRONMENT – location of Aliens and their symbionts (dungeons, underground nests, terraneous, mountainous or underwater isolates, and possibly aerospace objects). NESTS – underground bases and habitats of Aliens and their symbionts. Presumably connected by tunnels.

ALIEN CIVILIZATION – a community of Aliens and Anthropes, including their material and mental culture.

CONTACTEE is a person who had experience of communicating with representatives of an Alien civilization. ABDUCTION are kidnapping of people and livestock theft for different purposes: organ harvesting or for reproductive procedures. They could be returnable. ABDUCTANT – kidnapped.

MUTILATION – homicide involving blood and organ taking, is performed using tools or technical means. Usually it is made by UFO pilots.

CHUPACABRUS (sing=pl) are predators or vampires, presumably Alien's fighting cast. These include Chupacabra vampires, Jersey Devils, Moth-Men, etc. They does not use any tools when attacking.

PART 1
GIANT EUSOCIAL INSECTOIDS

REPORT-96

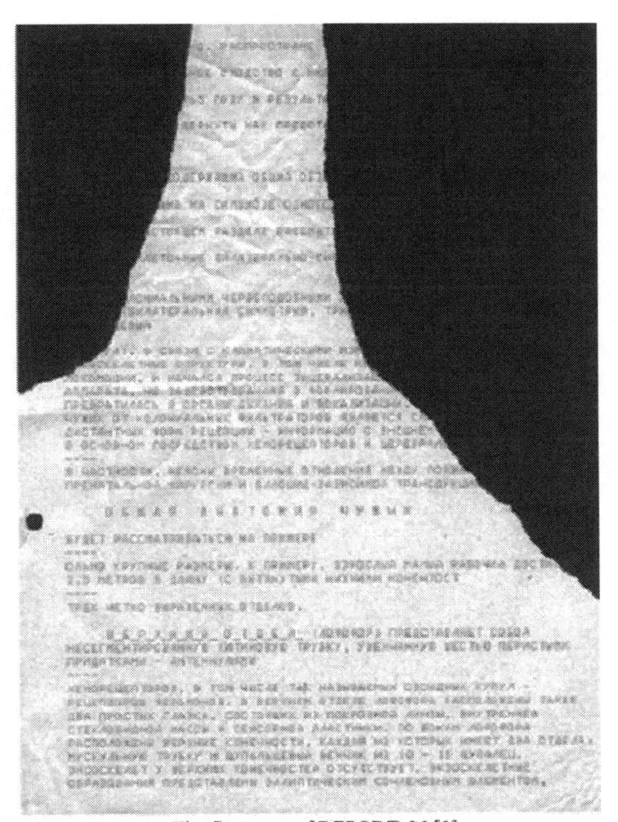

The first page of REPORT-96 [1]

In 1996, 27 pages of photocopies containing interesting text appeared on FidoNet [1] (APPENDIX 1). Author of a message, who wrote under the nickname «GVOZD», was disappeared soon, there was any possible to find him. Information presented on the dubious yellowed sheets could be believed or not. However, living creatures described resembled real nocturnal predators familiar to livestock breeders from many parts of the world under name of Chupacabra or Jersey Devil. Later, the same text, was named «REPORT-96», according to the first time of Internet publication. It was repeatedly reproduced on other sites in photocopies [2] and in printing text (see APPENDIX 1). For example, one of its early online editions appeared in November 2000 in a mailing list «Your world through the eyes of an Alien» [3].

REPORT-96 contained results of study of unknown cryptides. According to description, these were arthropods. However, it was surprised not only by their size, which was more than two meters. Some representatives of this species, including many casts, were intelligent, capable of communication, object manipulation and creative activity. Thusly they were carriers of qualities that are usually attributed to only one species – *Homo sapiens*.

Information of this kind always evokes natural distrust. However, REPORT-96 contained some forward-looking information. These discoveries were impossible to make «at the tip of the pen», even after processing a very large amount of special scientific information. Therefore, above deserved close attention.

The corners of the first text page were carelessly torn off. From rest of the lines it was possible to understand that it contains secret information that is not subject to disclosure. However, these sheets lacked signatures and details of an institution. It was obvious that the papers were not an official document. They were not a preparation of a scientific publication, since article presentation used to formalize other way. Perhaps these sheets were a synopsis or a draft of final report at a closed seminar for inducted listeners. Surely the text was consolidated. Such reports are generated from selected excerpts provided by laboratory leaders. In scientific institutes of former USSR, this was a common practice.

Ваш Мир глазами пришельца

Рассылка закрыта

Статистика

SUBSCRIBE.RU Выпуски Группы Подборки

Ваш Мир глазами пришельца 12 part

Служба Рассылок Subscribe.Ru проекта Citycat.Ru

[illegible] архив рассылки " Ваш мир глазами пришельца "

[illegible] - сайт галерей и изображений, где любой может открыть свои галереи.

Рассылка (Ваш мир глазами пришельца) 13 выпуск.

Сегодня очень любопытное творение, присланное читателем рассылки...Речь идет о документе описывающем "Чужих ". Судя по описанию, это не те "чужие" из одноименного фильма, а вполне гуманоидального вида (по происходящим внутри процессам). Пришельцу такие существа не знакомы, хотя довольно часто упоминается нечто похожее в описаниях различных экспериментов военных.

Что-то не видно, не слышно, глухо, как в лесу... Когда же всё-таки выйдет новый номер рассылки? Люди истосковались уже. Помнится вначале, был разговор о том, что Вы подведёте к тому, что же на самом деле "Ваш мир глазами пришельца". И где обещанные доводы? Лишь факты из жёлтой прессы. Надеюсь на продолжение рассылки. Знайте, что она нужна людям. Кто, как не добровольцы откроют глаза на мир всем жителям нашей планеты? Истина должна проявить себя в любом свете, как бы то ни было. Поэтому хочу послать Вам документ, найденный мною сегодня в Интернете, а Вы сами решите: истина это или нет. Сам я точно не могу выразить мнения об этом документе. Либо его написал "сдвинутый", либо это истина, но тогда насколько же зашли эти "вышестоящие" в своих экспериментах? Сколько можно скрывать? Хотя всей правды знать всё равно нельзя. Нельзя и в соображении безопасности страны, да что там - и Мира, и Планеты Земля...

Кто они, зачем явились сюда и где сейчас? ...а может быть здесь? Может быть, тот Вояджер сделал своё дело и донёс истину другим разумам, другим мирам и цивилизациям? А может Вояджер - это очередная ошибка человечества. ...Роковая ошибка! Но нет... так нельзя...

Истина где-то там, и пока она существует, будут существовать люди, живущие и дышащие той мечтой, мечтой добыть, во что бы это ни стало, добыть эту истину!!!

Документ вложен в сообщение с ошибками и пропущенными словами в первоначальном [illegible]

[illegible]

Общая анатомия чужих

Internet page "YOUR WORLD THROUGH THE EYES OF ALIENS" (Nov 2000) [2]

Photocopies were not done in Photoshop, as font on folds had some distortion. The letter outlines were slightly changed, but there were some differences in thickness of lines on fold, which was visible when enlarged. The sheets were torn one by one from the folder with round holders. The paper on which the text was printed looked old, although cheap paper can take a similar look in six months.

Insectoid head after dissection

The text was printed on a printer. In some places, side cuts of sheets are visible without traces of perforation or its breakage. Sometimes lines are skewed, as with a loose sheet feeder. Apparently, a dot matrix (needle) type printer was used for printing longtime. The letters were printed not in one stroke, but with an eight-pin print head, like on the Epson FX-80. Although scan quality was poor, the resolution was visible.

There were no hyphens except for those cases when it was present in the words. Sometimes they skipped up to a third of the line if a next long word did not fit even on letter. The author didn't really care about formatting and text integrity. Probably, the text was typed in a «LEXICON» program and not earlier than 1985, since this editor was created only that year. Thus, it turned out that the text was created between 1985 and 1996.

Be that as it may, the questions were raised not so much by the specific editions of the text of REPORT-96, as by its very content.

As to confirm reality of the existence of creatures similar to those described in REPORT-96, in 2019 intriguing photos appeared on social networks FACEBOOK. They show a head of insectoid, apparently partially dissected. At first glance, it does not seem very large, because of the hand with the scalpel in the foreground. However, on the left side of the picture you can see fingertips of prosector: they are commensurate with cryptid eye. Thus, the insect head can be commensurate with a human. This photo was exhibited by Russian ufologist Nikolai Subbotin.

POPULATION

According to REPORT-96, these cryptids were discovered in 1965. By the time the report was written, this project had been closed and research had been phased out, apparently as dangerous. The nature of danger was not specified.

According tradition, usually extraterrestrial being are named «aliens». Confirmation that the authors had them in mind was in almost every paragraph of text, where «aliens» were compared to «terrestrial multicellular organisms». At the same time, aliens allegedly descended from some kind of «worm-like bilaterally symmetric animals.» But most of terrestrial animals, both arthropods and chordates, descended from them.

Of course, REPORT-96 described organisms of terrestrial origin. This was evidenced by their cellular structure, chromosomal apparatus, and ability to oxygen of the air breath. These animals were clearly insects. However, I was surprised by their size, more than 2 meters. After all, the largest living insects are very small.

The external description of cryptids in REPORT-96 looked very stranges, however, their physiology, cytology and biochemistry were described in detail.

The Alien's community consisted of castes. Authors of REPORT-96 mentioned such representatives as «small worker», soldiers («fighting individuals»), and «upper caste individuals» or «reproductives». Since the members of the population differed not only functionally, but also physically, structure of their population could be called «eusocial» (really social). According to REPORT-96, Aliens lived in nests, which are a collection of some kind of underground structures. Apparently, there were many such nests, and their tenants could well distinguish between their own and strangers. It was about oxygen-poor underground isolates. The optimal oxygen content for Aliens was 13%. They easily tolerated high concentrations of carbon dioxide and humidity. However, Upper castes were rarely appeared on the surface, since breathing atmospheric air for twelve hours caused them poisoning.

The Upper caste individuals were engaged in intellectual activity, incl. and «religious ceremonies» (?!). In the «termite's mounds» there were also workers, who, in turn, were subdivided into podcasts. «Small Workers» were mentioned, therefore, must be «Large Workers» else. It is difficult to imagine their size, since the «small» ones with extended limbs reached 2.5 meters. These beings went out to «open space» at night and in dark. Perhaps «Small Workers» were a foragers-hunters, armed with special mouthpieces for attack. They avoided light, and when frightened, they showed a specific reaction, jumping up and to the side [4, 5].

Workers did other work in the nest, caring for eggs and larvae, as well as the «symbionts». Apparently, it is about some kind of cultivation or breeding. But it is more likely that it is animal husbandry and selection, because according to REPORT-96, «symbiotes recognize a host, who looks after them, by a smell of his pelvic glands».

Social lifestyle of Aliens and some of their physiological characteristics were very reminiscent of termites, whose society also consists of castes. Usually, a king, queen, workers, soldiers, larvae and eggs coexist in a termitary at the same time. In some species of termites, many reproductive pairs can live together, and castes can be divided into podcasts. However, specialized working castes are formed only in highly organized families, for example, *Rhinotermitidae* or *Termitidae*, and in primitive termites, their functions are performed by larvae.

Interestingly, some species of known termites are also engaged in something like agriculture and even use biotechnology [6, 7]. Often, termite mounds are also inhabited by simbiontes, representatives of other invertebrate species.

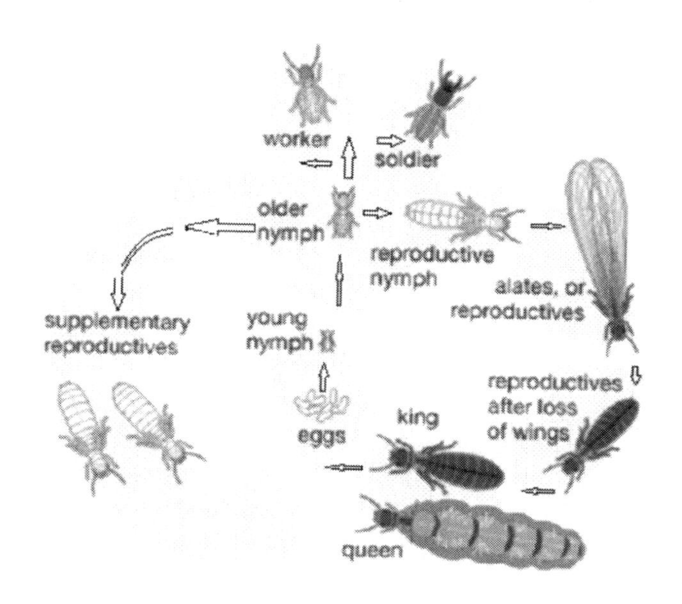

The structure of Termite colony

Reproduction of «Aliens» is carried out by pairs of Upper casts. Apparently there are many such pairs, they are diploid. Representatives of other castes are haploid. It has always been believed that not only Upper castes, but also other castes are diploid among termites [8]. Until recently, haploids were not found among termites, although their parthenogenesis has been known for a long time (since 1944) [9]. But in 2007 it was shown that in *Reticulitermes speratus*, there are many offspring, haploid casts, as a result of a special type of parthenogenesis. In these cases, haploid females and males (nymphoids and ergatoids) are born [10]. This article was published in 2007, much later the creation of REPORT-96.

There are also circumstances that allow ancestors of Aliens must be attributed to termites. For example, a strange phenomenon is observed in their cells, associated with the variability of karyotype: some small fragments, ring-shaped chromosomes are constantly present in their chromosome set. And although the authors of the report demonstrate ignorance, a similar phenomenon has long been noticed in male termites. During meiosis, they also undergo fragmentation and translocation of various parts of chromosomes, and formation of separate fragments and ring chromosomes. In particular, this phenomenon has been found in some species of African termites [11] and later in other species. The role of this phenomenon in formation of a termite's sociality has been discussed for a long time.

In addition, according to authors of REPORT-96, number of castes, nature of chromatin inheritance, as well as hereditary traits in «Aliens» is regulated by some mysterious factor of «allocyde-dependent transduction». It is allegedly foodborne and plays a significant role in the variability and composition of the population. It is under his influence that the karyotype undergoes constant changes and is «identical in only from one Upper castle pair generation». Perhaps this factor is a «tool», a vector for some kind of directed genetic modification, which Alien's carry out on themselves or symbionts. For a long time it was completely unclear what authors of REPORT-96 had in mind.

However, it turned out that a similar process was found in the termites of the species *Reticulitermes flavipes.* Under influence of some unknown nature factor an additional number of reproductive pairs are formed in the population of this species. Moreover, offspring of one pair have identical nuclear DNA, but has additional alleles in mitochondrial DNA. Perhaps this mysterious factor is associated with microbial symbionts transmitted through maternal line. *Information about a termite's factor similar to Alien's «allocyde-dependent transduction factor» was published in 2008 only* [12]. This is one example of forward information contained in REPORT-96.

ANATOMY AND PHYSIOLOGY OF ALIEN

EXTERNAL STRUCTURE

The features of the external structure of Aliens in REPORT-96 are not described in great detail. A speaker apparently used illustrations that are missing in a available text fragment. In addition, there are non-traditional names in REPORT-96. For example, the term «lophophore» usually refers to protostomes rather than arthropods [14]. The use of non-traditional terms may be due to secrecy or the fact that the initial stages of research on unusual organisms were carried out by specialists of a different profile (see below). However, the non-trivially named parts of Alien's bodie still corresponded to organs of real termites. For a better understanding, we will have to compare structure and function of these strange organisms with termites, as their closest known relatives.

Alien's populations include representatives of several castes, each of which has its own specific characteristics. In REPORT-96, external structure of Alien's body was presented on example of a Small Worker, a typical representative of one of castes. Judging by armament, Small Worker can be a Soldier. Interestingly, in the genus Macrotermes, soldiers (which in these termites are genetic females) come in two types: small and large. The former protect the workers during gathering of food or «repair work» on walls of the termite's mound, while latter guard chambers where a «queen», «king» and larvaes live [79].

The researchers divided the body of Alien into three conditional parts: «lophophore», «branchiophore», and «viscerotheca». It is quite obvious that behind these non-traditional names lie quite ordinary parts of a body for insects. Usually entomologists divide an insect body into head, thorax and abdomen. A thorax is divided into prothorax, mesothorax and metathorax. Such a division is also arbitrary, since these parts of insects bodies originated from different segments of a body of their distant ancestors, annelids. For example, an insect head usually consists of six fused segments. A termite even thorax and abdomen fused.

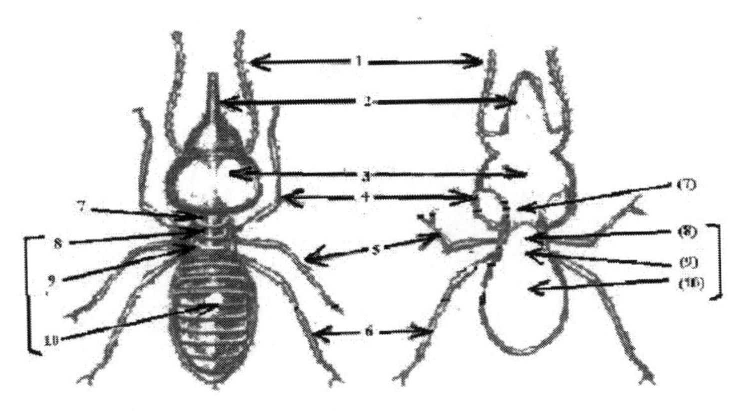

The relative position of body parts
of *Rhynchotermes* soldier and Alien

The head section of Alien's body contains processes corresponding to antennae (in REPORT-96 they are called lophophore limbs) and feathery antennules of insects. The main, unpaired process, «lophophore», is a wide hollow tube. The head section also contains paired compound eyes and simple ocellis. There are also a mouth and its specific processes (although authors of REPORT-96 erroneously attributed them to the branchiophore).

Branchiophore corresponds to an insect prothorax. Paired forelimbs grow from this department of insect body. They are laid in form paired germinal processes at larval stage. In Aliens, it is not limbs that form in corresponding places, but tubes of a pseudoxyryx (authors of REPORT-96 believed that lungs are formed here elso).

The lower part of Alien's body named «viscerotheca'. It is homologous to fused mesothorax, metathorax and abdomen of termites. In insects and Aliens, paired legs are formed from sections corresponding to mesothorax and metathorax. In termites, these are middle and rear legs, and in upright Aliens, they are upper and lower limbs respectively.

BODY COVERS, SENSORS AND LOCOMOTIONS ORGANS

Alien's body like all arthropods is covered with an external chitinous skeleton. Chitinous plates are concentrated in the upper layer of the exoskeleton. Chitin is secretion product of a myoepithelium, cell layer of which is located below. Plates are penetrated by thin channels containing cytoplasmic processes of myoepithelial cells. The integuments of termite's body are arranged in much the same way: they are formed by several cuticle layers, consisting of chitinous plates, under them there is a layer of hypodermal cells that secrete chitin. Among them are cells with long excretory ducts [13].

1 – lophophore, 2 – front yeys, 3 - rear eye; 4 - air bags; 5 – antennas; 6 - dissector fingers

Partially dissected Termitoid head: front and side wiev

Unlike termites, Aliens have an addition internal skeleton, it is a powerful chordal ring absolutely necessary for such large animal to maintain body shape and internal organs. They are attached to the chordal ring by mesenterys.

«Lophophore» is a head part of Alien's body, which is a «hollow tube». Apparently, it is homologous to a head process of soldiers of nosed termites of *Nasutitermitinae* family. Their nose-tube serves to spray caustic secrets that scare away enemies [79]. This massive formation is visible in a photo portrait of a giant insectoid. Authors of REPORT-96 wroted Alien's navigate and communicate using magnet echolocation (see below). So such hollow tube can play role of sonar, similar to dolphin echolocation organs.

According to REPORT-96, there are elso paired «lophophore limbs», which in their structure are more reminiscent of antennas of ordinary termites. They are flexible, because they are based on a hollow tube covered with chitinous «scales» with tentacles at the end. In termites, these are organs of contact chemoreception (taste) [13, 15]. In addition to the «lophophore upper limbs», some Aliens may have six «feathery antennae», antennules. These are hollow processes of the exoskeleton, which branche of the first, second and third orders extend. There are drops of viscous mucus on surface of third order branches. Termites also have similar antennules, they are smell organs. [13, 15, 16, 18, 19].

Alien's eyes, like termites, are complex, consisting of 15-16 thousand simple eyes. In addition to compound eyes, there are paired additional simple ocelli (osselli). A simple ocellus consists of a sensory plate of 30-40 light-sensitive neuroepithelial cells and 50-60 cells located between them that produce the vitreous substance. The eye is enclosed in a connective tissue capsule. Its intercellular substance contains grains of a dark pigment of an unknown chemical nature.

The resolution of compound eyes is high, but less than that of humans. Aliens have binocular vision and the ability to judge the distance to objects, but it is imperfect. Aliens are able to identify the stereoisomers of certain organic compounds. A similar ability of insects is also known. It was discovered in honey bees by K. Frisch in 1948. Later it turned out that all insects and their free-living larvae, as well as arachnids and crustaceans, are able to orient themselves in the plane of polarization of the rays. Like termites, some Alien's castes may have compound eyes of varying sizes or none at all, such as workers or soldiers.

In Alien's head section, close to the sensory brain, there are paired magnetosensor cavities, the walls of which are formed by the muscle layer, and the volume is filled with colorless «magnetolymph». A dozen and a half spherical bodies are suspended in it, each of which contains crystals of iron salts, surrounded by nervous tissue. Apparently, the function of this organ during research was not sufficiently studied, but an opinion was formed that it are magnetosensor and are used for orientation in space and for information exchange «in three-dimensional dynamic images.» This magneto sencitive organ compensate for Alien's imperfect vision.

According to REPORT-96 authors, this organ is a source and detector of an alternating magnetic field that changes its configuration under influence of surrounding objects. Apparently, the magnetically sensitive organ of Aliens should work on the principle of a locator. It can be assumed that the hollow lophophore may play the role of dolphin sonar.

Apparently, the magnetosensor organs of Aliens did not arise out of nowhere, but developed based on similar abilities of their insect ancestors. Evidently termites elso have some organs performing similar function. Like many other insects, when stopped, position their body axis is parallel or perpendicular to the magnetic field lines. Termite's queens orient their body axis along a north-south line. Perhaps in the dark, where navigation by vision is not possible, worker termites are able to maintain direction with the help of a magnetic field receptor. However, location these receptors and how magnetic field intensity is perceived in insects is not clear yet.

The section in which the perioral organs of Aliens are located was called the «branchiophore» by researchers, although in insects it is traditionally referred to as the head section. The mouthparts of Aliens have a structure typical of termites, which depends on caste specialization.

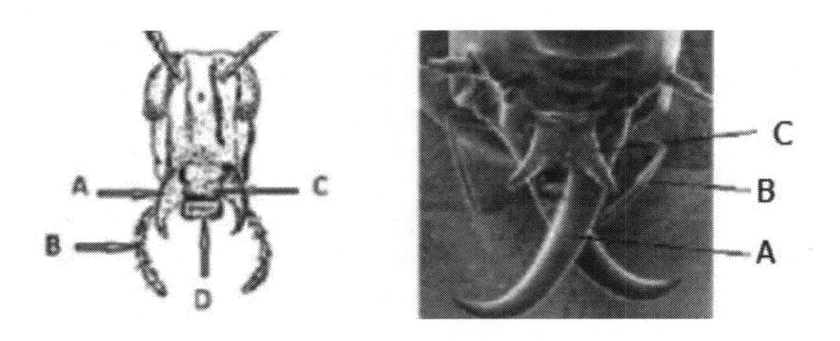

Near-mouth organs of common termite (left)
and *Cubitermes* soldier (right):
A – superior mandible, B – grasping palp, C – upper lip (Labrum);
D – lower lip (Labium) (it is not visible in the right picture)

Alien's Small Workers oral apparatus is very complex and corresponds to their activity as a predator and forager. Apparently, other castes do not have such a complex mouth apparatus. Other castes feed on semi-digested food brought by foragers. (Probably, the Alien in the above photo does not belong to a forager caste, since its mouth opening is just a gap, not surrounded by any processes). However, the text contains a shocking reference to cannibalism practiced by the upper castes.

The mouth opening of Small Worker is surrounded by a fibromuscular membrane, homologous to termite's lips: upper lip (Labrum) and lower lip (Labium), but closed in a ring. Sharp hook-shaped needles stick out from there. These are modified paired upper (mandibulae) and lower (maxillae) jaws of insects. With its mandibles, Small Worker pierces a body of a victim and powerful circular perioral muscle provide negative pressure and suction of animal blood [22]. Length of hook-shaped needles reaches 25 cm.

Typically, termite's jaws are relatively shorter and their shape depends on nature of their diet. Although most termites have gnawing mouthparts, some lines of the family *Termitidae* have developed a crushing bite [20]. They have grasping palps, hook-shaped (sickle-shaped) upper (mandibles) and lower jaws (maxilla). They are covered with webbed lips, upper and lower

[79]. Most of all, the jaws of Alien's Small Worker just resemble saber-shaped mandibles of *Cubitermes* soldiers. This termite genus is belonging to the highly developed families Termitides.

Near the mouth, in the «branchiophore» of Aliens homologous to the prothorax of insects, there are paired bubbles of lungs and a psevdoxirixs. These formations, resembling flaps of tissue, consist of branched tubules. In the photo provided by N. Subbotin, a large insectoid shows bags of pseudocyrix and lungs. Apparently, after the autopsy of the body, they were preserved only on the right side [27]. In their structure, these formations are very reminiscent of the larval gills of aquatic insects. The lungs are a respiratory organ of Aliens. Gas exchange with blood occurs through their walls. They are filled passively, and exhalation forcibly drives the air mixture, saturating the blood. Unlike lungs pseudocyrix tubesare bigger diameter and are closed from on end. It is a paired vocalosation organ of Aliens.

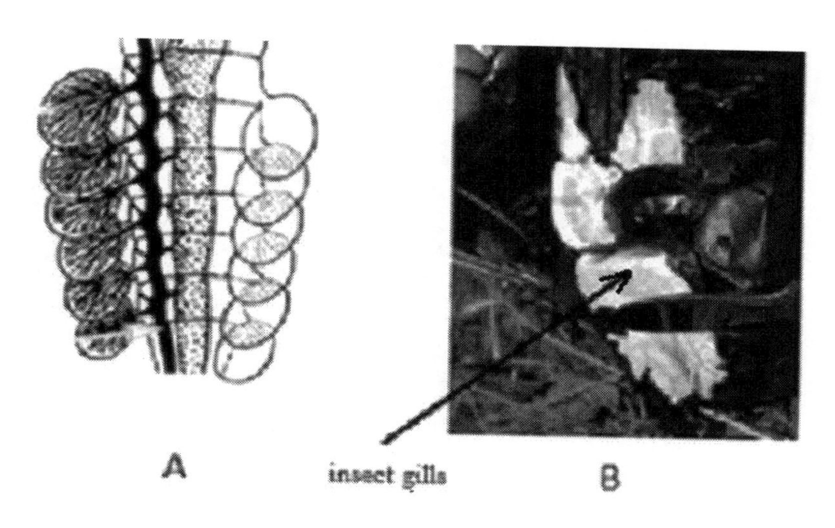

Larval gills of stoneflies (A) and Alien's lung sacs (B)

May be this paired bubbles develop from paranatal processes in the larval stage. As a rule, additional processes disappear at later stages of development. They form limbs. As a rule, extra processes disappear in later stages. This was also noted Fritz

Müller, who studied the development of South American termite larvae in 1878 [25, 26]. However, additional processes do not disappear in all insects. For example, they gives rise to a larval gills of aquatic insects. Many insects of the Upper Carboniferous had paranatal lobes on the prothorax corresponding to the gills of their crustacean ancestors, hawever larval gills are preserved only in aquatic insects [23].

Authors of REPORT-96 believed that lungs and pseudoxyryx of Aliens are formed in place of their forelimbs (without directly naming, apparently for reasons of secrecy, what they mean insects). However it is highly probable that only Alien's lungs are formed from prenatal processes homologous to larval gills, while pseudocyrix is formed from prenatal processes homologous to forelimbs of insects. In any case, the internal structure of the tubes of pseudoxyryx is very reminiscent of acustic or typanic organs of grasshoppers, also located in forelimbs [28, 32]. The inner surface of pseudocyrix tubes is equipped with chitinous ridges and hooks, due to the friction of which sounds are produced. Sometimes they resemble grinding or buzzing [29-31]. The tympanic organs of termites also form on the forelimbs. They make sounds by rubbing against the wall of the prothorax.

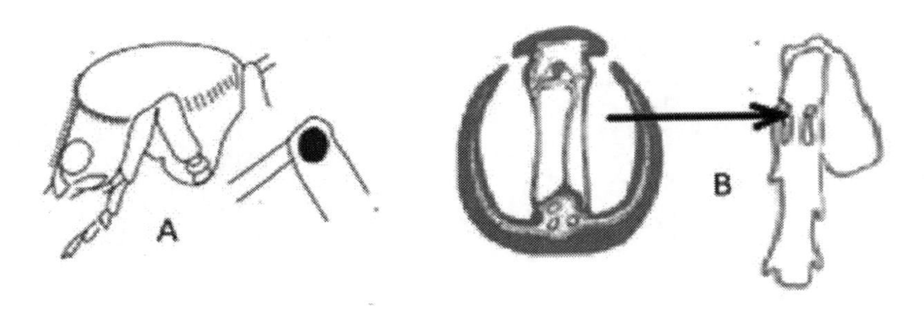

Termite tympanic organs: corrugated surfaces on prothorax and forelimb (A), acoustic organ in forelimb of grasshopper (B)

Embryonic processes, homologous to the forelimbs of the prothorax of termites, in Aliens turn into a paired vocalization organ – pseudoxyryx. These are bags made of tubes, the inner surface of which is equipped with chitinous ridges and hooks, due to the friction of which sounds are produced. The structures of pseudocyrix are very similar to the acoustic organ of grasshoppers (close relatives of termites). Its cavities are located in the forelimbs [28, 32]. Apparently, this is the origin of the vocalization organs of Aliens. Indeed, some witnesses have claimed that their sounds are like to a grinding, chirping or buzzing [29-31].

Apparently, due to transformation of the forelimbs, Aliens do not have a localized hearing organ [13]. There are vesicular sensilla, small cavities in the exoskeleton filled with air. Cavity outer wall is thin and begins to stir under a resonant frequency. The inner wall of the cavity is composed of softer chitin, riddled with nerve endings. Resonant frequencies are different for sensilla located in different parts of the body, which allows you to perceive sound and vibration in a wide range of frequencies. Alien's touch organs (perception of pressure, sound vibration transmitted through a solid substrate) are structured in much the same way. These are vesicles deepened into a cuticle layer, from which a hair sticks out. Absolutely the same organs are distributed in the surface layers of the cuticle of termites [13].

The «viscerotheca» is the rear part of the Alien's body. It does not have a pronounced division into segments and is a merged section of the mesothorax, metathorax and abdomen. Paired limbs, «upper» and «lower», are attached to the viscerotheca upper part. Thus, the animal must be upright, otherwise theirs limbs would be called «anterior» and «posterior». Nonetheless the upper and lower limds are homologous to the middle and hind limbs of insects, respectively. Obviously the upper limds play role of «hands», so necessary for a rational being to manipulate objects. They are homologous to the middle pair of legs of insects, derived from mesothorax. What is their structure is unclear, since the part of REPORT-96 text is missing in this place [33, 34].

However, other sources contain information about the upper limbs of Aliens. According to some witnesses, the hand resembles a broom handle. The «hand» is wider than a human, but not much. Two pointed «fingers» without nails, they grow directly from the «wrist». One is longer, motionless, the second is movable [19]. This description is very reminiscent of cancer claws. Soft appendages elso may be here, because sometimes eyewitnesses saw something resembling fingers between pincers [31]. Little more is known about the lower Alien's limbs from REPORT-96. They are homologous to hind limbs of common insects derived from metathorax. Their structure is typical of legs of cockroaches or termites. They consist of tubes whose muscles run inside, a coxa, thigh and tubia. Between them are two articulated elements. A «foot» consists of a heel extension and powerful spurs. Apparently, ther prints are similar to cancer claws or a cloven hoof [35].

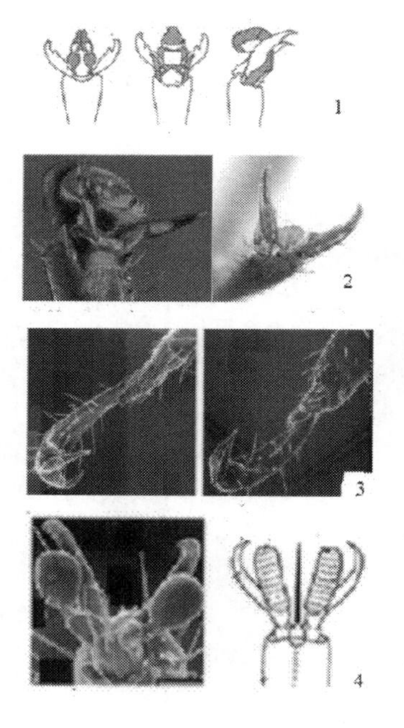

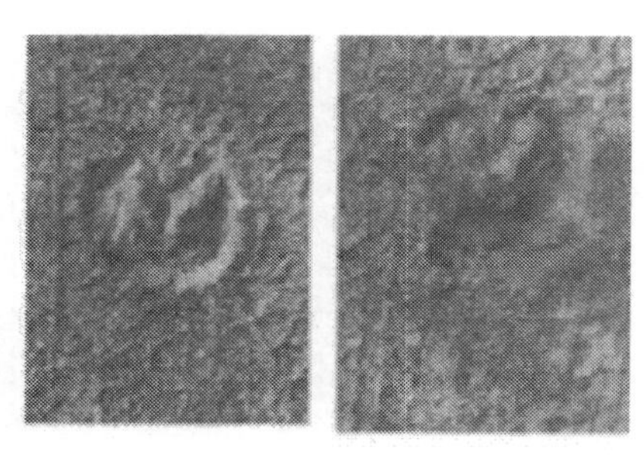

Insect and Alien paws: 1- diagram of base Insect model paw: (top view, side view, and bottom view);
2 – microphotography of a cockroach paw - claws with a central process, aronium (bottom view and side view);
3- microphotography of Cubitermes termite paw– claws (two different side view); 4 - microphotography and diagram of bug paw - claws with central emodium and pulvills (bottom view); 5 - paw print of Chupacabrus from Gryady village (Ukraine); 6 - paw print of Chupacabrus from Chechersk (Belarus).

Insect and Alien's paws

Termite's limbs are also very similar to this description. The structure of their foot, just like that of other cockroaches, repeats a «basic model» of claws of crustaceans, their distant ancestors. It contains two large pincers. However, cockroaches and termites may have additional «arolium» appendages, which can be clearly seen in micrographs of cockroach paws. In addition, many insects have evolved soft outgrowths, pulvilles, which allow insects to attach themselves to smooth surfaces.

There is no mention of wings in the accessible fragment of REPORT-96. However, only one working caste is described there. It is possible that adults are winged, and as are termites during the nesting period. Usually larvae do not have wings, while nymphs can have only rudimentary ones [36].

The posterior part of the viscerotheca is homologous to ordinary insects abdomen and contains the main vital organs of Aliens [13].

EXOCRINE GLANDS AND PHEROMONES

Pheromones and other exretions of termite's exocrine glands are most often steroid substances and are transmit through smell or food. They regulate behavior of members of population, stimulating or suppressing their development [56, 57]. Alien's pheromones are very similar. For example Alien's subopharyngeal gland secretes a polyisoprene pheromone that causes workers to regurgitate feeding, like to secret of salivary glands of termites. Alien's coxal glands secretes a pheromones that are recognize by others nest members and symbionts. The smell of Alien's pelvic secret causes a stupor in worker cast and aggressive attack of soldiers. Apparently, it is analogous of a specific pheromone of termite's thorax gland [13]. Aliens also have a «status gland», which secret indicates belonging to the dominant caste. Possibly, it is an analogue of adjacent termite's body secrets, which change a size of female termitids [13].

The secrets of Alien's Intrasiryngeal Glands cause immediate reactions from other population members. Just getting in the olfactory organs, they are quickly transported along olfactory fibers into a brain capsule liquor. After interacting with the receptors of neurosecretory cells of an intermediate brain, specific oligopeptides are synthesized. They cause a certain emotional and involuntary automatic reaction, different for each caste.

Alien's behavior regulation is apparently the result of evolutionary development of a similar system of termites. Their pheromones cause an immediate reaction of alarm, flight or aggression depending on caste. However, Alien's system, apparently, has reached a higher degree of development and, may be this way this communication was named a «pheromone component of the speech» of Aliens.

NUTRITION AND DIGESTION SYSTEM

The most important for understanding the direction of evolution and for the taxonomy of termites is study of their digestive system and intestinal microflora. However, in Aliens, entire lumen of digestive tract, with the exception of some specialized forms, does not have a resident microflora. Apparently, this is due to the specificity of nutrition, which is very different from that characteristic of any species of termites, in which the main food source is plant residues containing lignin and cellulose. To digest them, termites need symbiotic bacteria, flagellates or micromycetes. However, Aliens have not resident microflora in the intestines, so there is reason to believe that it caused by switching from saprophytic to a predatory way of Alien's ancestors feeding [5].

Alien's Workers «care for symbionts» in nests. Apparently, we are talking about symbionts living outside of bodies of Aliens, but inside the nest, and we can talk about cultivation or selection of pets. It is interesting that some species of termites belonging to the Termitidae family are also engaged in cultivation and even directed selection of micromycetes [6, 7]. In termite's nests, you can often also find a lot of other living creatures so-called «thermophiles», belonged to another kind of insects, worms, and even small birds.

Apparently, eating symbionts plays an important role in the diet of Aliens. With food, they also receive the «allocide-dependent transduction factor». In addition, authors of REPORT-96, while silent about the main source of nutrition, assumed that Aliens have some protein deficiency, since thay eat organic remains, insects and even using cannibalism (among Upper castes). In any case, chitinases were present among the digestive enzymes [37, 38]. Probably, such assumptions were made by analogy with termites, which, due to protein deficiency, can eat the remains of insects and even their sick fellow tribesmen.

There is an indication in REPORT-96 that Aliens are capable of assimilating proteins of both animal and plant origin, and that phenylalanine is an essential amino acid for them. Essential ions for Aliens are calcium, phosphate. In addition, Aliens need trace elements of copper, manganese, zinc and iron. The authors of the report mentioned in passing that Aliens get all this from food at the expense of symbionts. At the same time, they kept silent about the source of iron, but Aliens need a large amount of it. It can be assumed, given a clearly predatory way of feeding and the mention in REPORT-96 of «grasping-devouring» perioral appendages, that a main source of organic matter and iron for Aliens is the blood of warm-blooded animals. It is a good and readily available source of ferrous and ferric iron in the form of hemoglobin and met-hemoglobin. It is part of their chitinous integuments and magnetosensor organs. May be vampirism of Aliens was not discussed even in a highly classified research society.

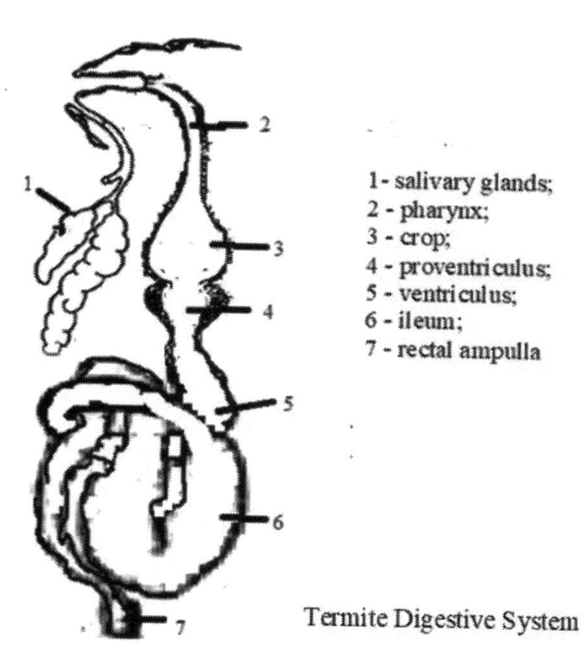

Termite Digestive System

Despite a fact that structure of Alien's digestive organs has a lot in common with termites, the transition to a predatory way of feeding has made certain adjustments. Small Worker's mouth apparatus, originally a gnawing type, is redesigned (see above) to pierce tissue with long hook needles (25 cm) and suck out fluid. The secrets of the myoepithelium of the Aliens digestive tract contain enzymes with a pH optimum in the slightly alkaline and neutral regions, which can be active in the victim's blood plasma at pH 8.4.

If tropholaxis is characteristic of most termites, due to a need to transfer the intraintestinal microflora, then Aliens, who do not have a resident intraintestinal microflora, are unlikely to have tropholaxis of this type. Nevertheless, to feed larvaes and members of other castes, Small Worker forager regurgitates a gruel-like semi-liquid mass of a semi-digested product, which does not require further processing.

Mouth organs of aliens and termites are very similar (see above). If we compare the digestive system of aliens and termites, we see another significant similarity, although the terminology used to describe may not be the same. The digestive tract of termites and aliens begins with the pharynx. The pharynx is followed by a goiter (crop) and a muscular stomach (proventriculus), where food is chopped. The Small Worker's forager stomach holds about 20 liters. The cardinal sphincter separates the proventriculus from the small intestine (ventriculus), which is covered with muscular folds. It contains many glands that release hydrolases. Their release occurs by detaching the vacuoles of secreting cells. The pyloric valve (Alien's «intestinal sphincter») separates it from the next department large intestine, hind gut. Here, water is reabsorbed and feces are formed. Phagocytes loaded with microorganisms and waste phosphorus compounds are thrown here. This section closes the pyloric valve (Alien's «anal sphincter»). The last section of the digestive tract is the rectal ampulla (Alien's, the «anus»). The ducts of the malpighian vessels flow here (Alien's the «green gland»). The secret of the gland is mixed with the feces.

The length of termite's digestive tract is about 5 times of length of its body, although this ratio largely depends on the lifestyle of the termite and the way it is fed. For Aliens leading a predatory lifestyle, this attitude may be different.

Aliens, like termites, have a fat body («fat pad'). It is analogous to the liver of vertebrates. Theirs cells contain clearly visible vacuoles, they are lying between the internal organs. In fat pad cells the accumulation of lipids, sugars and amino acids and biosynthesis of important blood coagulation factors occur.

BLOOD CIRCULATORY SYSTEM and RESPIRATION

Annelids are distant ancestors of insects. Their circulatory system was closed. Blood circulated through large vessels, which fanned out, passing into capillaries. Capillaries ensured the transfer of nutrients and oxygen to the organs of the body. Then the capillaries were again assembled into large vessels. Later in the process of evolution, arthropods appeared. At the same time, their circulatory system was partially reduced. It is partly open into the intercellular space in crustaceans, and almost completely open in insects. Fluid enters the body cavity from large vessels located on the dorsal side of the body, and then is collected back into the vessels. Therefore, this fluid is called «hemolymph» (blood and lymph). Insect hemolymph transports the products of food digestion to metabolically active organs and tissues. In this case, oxygen mainly passes through the branched trachea [40, 41].

A circulatory system of Aliens is closed. Apparently, its reduction was not complete and for some reason closed again, like in their distant ancestors of worms. Blood and lymph (interstitial fluid) are spatially separated everywhere. Two large annular vessels and two connecting ducts with pulsating muscle fibers embedded in their walls are the main elements of a Aliens' circulatory system. Vessels supplying internal organs: pharynx, stomach, intestines, green gland and sex glands (in Upper castes) depart from a lower ring. From the internal organs, vessels gather into veins that flow into the ascending stream. Paired pulmonary arteries and blood vessels supplying the brain, pseudokyryx and lophophore depart from the upper ring. The Alien's heart consists of two accumulations of blood smooth muscle cells in the walls of the arterial and venous connecting ducts. They contract in opposite phase. Contractions occur autonomously, but their frequency can be regulated by a splanchnic nerve as well as by humoral factors. The blood circulating in the Alien's circulatory system is mixed. The degree of venous and arterial mixing in blood is regulated by the contraction of smooth muscles embedded in the vessel walls of upper and lower rings, under the influence of nerve and humoral impulses, which increases the adaptive capacity of the body.

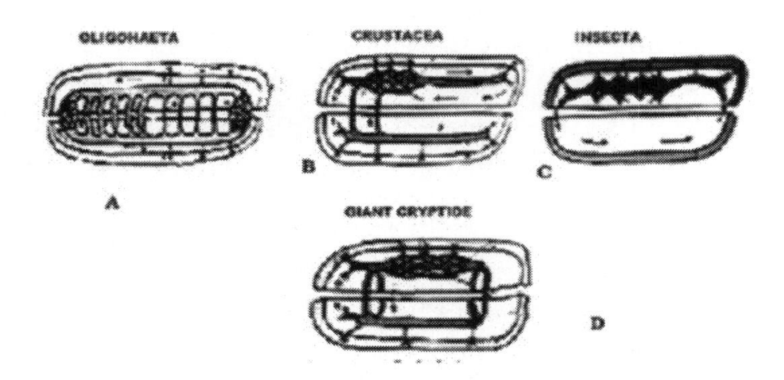

Lymph and blood circulation systems in annelids (A), crustaceans (B), insects (C) and Aliens (D)

Thus, aliens have 2 circles of blood circulation, like vertebrates. However, details of vessel structure significantly differ and rather resemble to structure of insect vessels: each vessel is surrounded by clusters of circulating smooth muscle cells. How could a closed circulatory system have phylogenetically evolved from an open circulatory system? Such a process is fairly well known in evolution, for example, a in oligochaete and many polychaete open circulatory systems had turned into a closed types [42]. Development of Alien closed circulatory system was closely connected with appearance and parallel development of a «lungs» from branchiophore prosess. Modern termites have not analogues of these organs (but some species of termites have their prenatal rudiments). Gas exchange with blood occurs through their trabecular walls.

Alien blood is an opalescent liquid that, depending on oxygen saturation, it may be a blue or greenish tint. Its color definition by «alienocyanin», that is a specific oxygen carrier. Obviously, this name was composed of two words «alien» (alien) and «cyanine». It is quite obvious that we are talking about hemocyanin. Hemocyanin is a copper-containing respiratory pigment from a group of chemoproteins. Hemocyanin is found in many groups of arthropods, horseshoe crabs, crustaceans, arachnids, and centipedes.

However, until some time it was not found in insects. It was believed that this chemoproteid lost ability to bind copper in course of evolution, transformed to storage protein, hexamerin. It has been shown that crustacean hemocyanin (*Crustacea*) and insect hexamerin (*Insecta*) have a common origin. Moreover, the time of divergence of hemocyanin and hexamerine seemed much older than the divergence of the classes themselves [43]. It was assumed that insects do without a special oxygen transporter protein due to multitude of branched tracheoles and a certain amount of oxygen dissolved directly in the hemolymph. However, absence of a specific oxygen-carrying protein, apparently, was one of some limiting factors preventing size increasing of these animals.

Therefore, it would seem strange that Aliens, descending from termite-like ancestors have hemocyanin. However, ability to sintese it had not been completely lost by some insects. Hemocyanin was found in hemolymph of larvae, one of a species of stoneflies Perla marginata [44]. Later, hemocyanin was also found in another species Perla grandis [45]. It should be noted that stoneflies and termites, are taxonomic close because they belong to the same subcohort Polyneoptera.

Even narrow specialists before 2004 did not know about a possibility of existence of insect hemocyanin and its moleculare structure. As it turned out in 2004, the stoneflie hemocyanin consists of two subunits, each of which contains two copper atoms. Surprisingly, the authors of REPORT-96, written a decade and a half before this discovery, accurately described what the subunit structure of Alien «alienocyanin» is: it also contains two subunits and four copper atoms. Like a stoneflie hemocyanin, it is also transported in a state adsorbed on hemocytes. To guess or predict these specific features is not possible because other classes of invertebrates, hemocyanin has a different subunit structure. Hemocyanin of arthropods (cheliceral) have been a 6-unit structure (540 – 420 million years ago). Later evolution of arthropods formed hemocyanin multimers (2 x 6, 4 x 6, 8 x 6). The hemocyanin of crustacean, insect ancestors, was a monomer or dimer of 6 subunits (1 x 6, 2 x 6) [46, 47]. Thus, the traits of Alien hemocyanin could not be known from any source, nor could they be «invented». This information could be obtained only as a result of *direct experimental research.*

Analyzing the description of blood physiology, one cannot ignore such an important issue as hemostasis (blood property to keep liquid state and to form a clot if a vessel wall is damaged). Formed elements and blood coagulation factors have main role in the formation of a clot and in its later lysis. At the damaging site of the vascular wall is a result of the release of the «thermostable coagulation activator», that is starter of cascade enzymatic reactions. Apparently, this factor has a lipoprotein nature (since it is cleaved by «plasma lipase»). This one and others coagulation factors, are biosyntezed in a fat body. The reverse process, clot cleavage, occurs when cell damage stops and the concentration of thermostable coagulation activator drops. The clot is colonized by fibrocytes and endomesothelial cells, capable to phagocytosis, they lyse it, forming cavities.

Insect hemolymphostasis is similarly maintained. Coagulation occurs due to coagulocytes, which accumulate in places of damage to the body, form a plug. Just like aliens, insects also have an alternative hemolymph clotting pathway. Cascade activation process proteolytic enzymes are elso synthesized in fatty body cells. The negatively charged lipids (similar to Alien's lipoprotein thermostable factor) have certain role in coagulation, they are secreted by hemocytes.. The clot is cleaved according to the same scheme as in Aliens, by phagocytosis with participation of hemocytes [48].

It can be concluded that despite some differences in structure of circulation systems in insects and Aliens, they main basic functions are very similars. Differences are Aliens closed circulatory system, presence of lungs and an oxygen carrier (hemocyanin). They developed during the evolution and are compensatory adaptations to an environment with low oxygen level.

OSMOSIS, TRANSMEMBRANE TRANSPORT, and EXCRETORY SYSTEM

Alien cells are much larger than those of other insects. Their sizes can reach 100 microns. Aliens have a completely unique structure of membranes and ion channels. Cell membranes have an unusual structure: on top of lipoprotein bilayer and glycocalyx there is a layer of sulfolipid granules, which is interrupted only at sites of intercellular contacts.

Sulfolipids are sulfate derivatives of cerebrosides, in the molecules of which sulfate is attached to third hydroxyl of galactose. They are strong anions and are involved in cations transport through membranes of nerve cells and fibers, and thus are needed for normal electrical activity of a nervous system, providing conditions for membrane polarization. However, sulfolipids impose restrictions on membrane mobility. In Aliens, sulfolipids are contained not only in nerve cell membranes, but also in all somatic cells. Probably, in evolution process, Alien's ancestors were in conditions that required minimal surface cell area and restriction of transmembrane transport.

There is a hypothesis according to which first prokaryotic organisms had appeared in sodium-free environment. Later, plasma membranes and mechanism for constant removal of sodium ions appeared as an adaptation for living in «hostile» sodium environment [50]. Multicellular organisms have at least two liquid phases: potassium-containing intracellular and sodium-containing extracellular. Both liquids have the same total concentration of osmotically active substances and a similar value of osmotic pressure. The difference in concentrations of potassium and sodium ions on both sides of cell membrane is provided by active transport of these ions with the expenditure of energy and participation molecular «potassium-sodium pumps». This is true for most animals.

However, there are some exceptions. These include, in particular, insects. High osmotic pressure of potassium inside their cells is compensated by high content of disacharides and polyatomic alcohols in extracellular liquid (hemolymph). The Alien osmotic balance between intra- and extracellular liquids are maintained the same way. The main intracellular cation of Aliens is potassium, and intercellular antagonist, like to insect, is organic matter (dimethylamine oxide). However, in comparison with insects, this evolutionary direction in Aliens has gone even further. Sodium ion is present in insect extracellular environment at least in small quantities. But there is any sodium in extracellular liquid of Aliens. Perhaps this is why Alien cells are covered with an insulating sulfolipid layers, their cells have not membrane pores. Cells communicate with each other directly through intercellular ducts.

It should be noted that *description of mechanism for maintaining the osmotic potential in Aliens does not contradict the noted structural features of their membranes.* Excessive release of potassium from their cells leads to difficulty in the transmission of excitation to a central nervous system and muscle contraction of Aliens. The same disturbances are observed in insects. They increase the permeability of membranes with excessive release of potassium ion to the outside. This happens under influence of insecticides [24].

The final product of nitrogen metabolism of Aliens, like other ordinary insects, is uric acid (in contrast to mammalias that eject nitrogen forming urea). The biosynthesis system of uric acid serves to bind toxic products of nitrogen metabolism (ammonia) and appeared due to a limited water balance, when excretion products are excreted with a minimum amount of water or in solid form. Just like all insects, uric acid is excreted from the Alien's body in a concentrated, almost solid form. The main excretion organs of Aliens are called a «green gland». (This name may have been chosen by analogy of green glands of crustaceans. However, there are not a homologous body parts). This paired organ is homologous to Malpighian vessels of insects. The green glands secrete mucus containing a polysaccharide. This is universal ion exchanger, it absorb a metabolic products, excess salts and toxins. In the rectal intestine, these substances are mixed with feces. Product excretion is colored by bright green. (Ufologists, discovering such smelly substance, are very happy, and try to guess what it is!)

An additional function of *excretion* from Alien's body is performed by blood phagocytes, similar to how it occurs in termites (in them this function is performed by hemocytes of hemolymph). Filled with waste phosphorus compounds, these cells enter in the space of the large intestine and are excreted with the feces. By the way, due to this feature, termites play a significant role in the enrichment of the soil with phosphorus [51].

CELLULAR RESPIRATION and SUGAR OXIDATION

Usually, cellular respiration in eukaryotic cells occurs in mitochondrias. However, they are absent in Alien cells. The authors of Report-96 write about «reticulated bodies». This reticle filled entire cytoplasm, its finger-like processes were drawn to organelles that actively consume energy. Despite membrane structure of reticular bodies was not entirely clear, there was not doubt of electron transport chains existence and of oxidative phosphorylation process, because histological reaction to oxidative phosphorylation enzymes of stained the network in dark blue color. Thus, the reticle performs some functions inherent in ordinary mitochondria. An electron donor was reduced form of a quinoid coenzyme, and final acceptor was molecular oxygen. When one electron pair was transferred from a quinoid coenzyme to oxygen, five pairs of protons were released into cellular space.

There is reason to believe that similar studies of Aliens were carried out in the United States. In a recently declassified report Q94-109 [77] (see below), as in REPORT-96, it was noted that Alien cells contain modified small mitochondria immobilized on the endoplasmic reticulum network. Apparently, Alien mitochondrias are greatly reduced. (Probably American researchers used not onle light microscopy, but electron microscopy else).

Thus normal electron transfer occurs in the mitochondrial reticulum, similar to all eukaryotic mitochondria respiratory chain transferring electron from reduced nicotine amide adenine dinucleotide (NADH) or reduced ubiquinone to molecular oxygen. Unlike eukaryotic cells, where mitochondrial membranes contain proton-dependent ATP synthetase and mane conjugated phosphorylation product is ATP (adenosine tri-phosphate), the same role in Alien cells play other macroerg GTP (guanidine tri-phosphate) because their mitochondrial reticular membranes contain proton-dependent GTP synthetase.

According to American researchers high role of GTP in Alien metabolism is also due to important role of Arf GAP proteins, which bind GTP and induce its hydrolysis. But *during during the period of REPORT-96 creation, this fact could not be known to researchers, since Arf GAP were discovered only in 1994 [77, 82]. Thus, the Sechenov's researchers only reflected what they learned from practical research and explanation of their results was received later.*

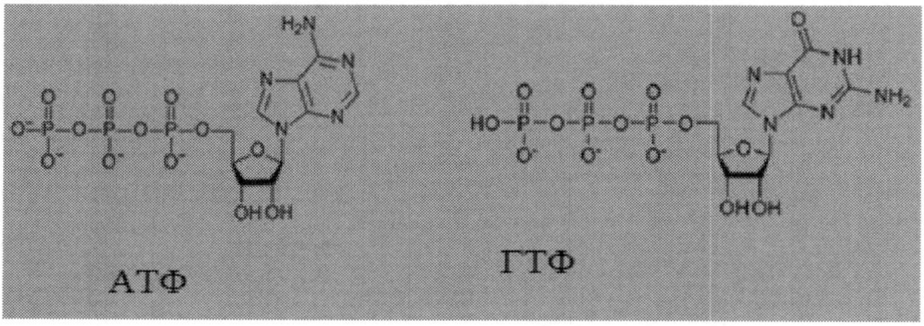

АТФ ГТФ

However Alien cellular respiration systems are very unusual. REPORT-96 noted that formation of oxaloacetic acid occurs through «bi-carbon compounds». Researchers found this aspect of cellular biochemistry «very strange». Indeed, in a Krebs cycle, which usually takes place in the mitochondria stroma, oxaloacetic acid appears as an intermediate, but its precursors are malic (4 C) and fumaric acids (4 C). Additional source of oxaloacetic acid is reaction between carbon dioxide (1 C) and pyruvic (3 C). These mane two pathways for formation of oxalic acid are universal and usually pass in eukaryote mitochondria. There are some lesser known biochemistry reactions with formation of oxaloacetic acid. One of them passes in mesophyll cells of oil crop plants, this is a reaction between bicarbonate (1 C) and phosphoenolpyruvic acid (3 C). Some microorganisms go through an Arnon cycle (reverse Krebs cycle), there oxaloacetic acid is a product of the deacelation of citric acid (6 C). All these reactions can not be mentioned as a specific pathway for formation of oxaloacetic acid from «two-carbon» fragments. However, there is another process, so-called glyoxylate cycle, which also forms oxaloacetic acid. It was first described by Kornberg and Krebs in 1957 [52]. Total reaction of the glyoxylate cycle is:

The acetyl (2 C) formed in process of hydrolysis of fatty acids, bound to coenzyme A, reacts with glyoxylic acid (2 C), further forming four- and six-carbon acids. The terminal glyoxylic acid (2C) is again involved in cycle. The key reaction of this cycle, catalyzed by malate synthase, involves to oxaloacetic acid formation from two bicarbon compounds: glyoxylic acid (2C) and acetyl (2C, it is bound on coenzyme A). Then two 2C-components forme four-carbon malic acid (4C) with is converted to oxaloacetic acids.

HCCOOH + CH$_3$C~SKoA → O–CCOOH / H$_2$CCOOH

glyoxylate acetyl-CoA oxaloacetate

Glyoxylate cycle is widespread in bacterias and micromycetes and allows them to survive even where there is no other carbon source besides acetic acid CH_3COOH. The glyoxylate cycle was elso found in gluconeogenesis of oil plants. However *intil 2005 this enzyme system was completely unknown in animal cells*. At first it was discovered in nematodes. Later, a specific genes of glyoxylate cycle, malate synthase and isocitrate lyase, were found in different groups of animals, including insectes [53, 54]. So, *REPORT-96 autors could not known about «synthesize oxaloacetic acid from two-carbon fragments» and this Alien's feature discovered in practical studies seemed unusual to them.*

It is very likely that oxidation of sugars in Aliens occurs not in reduced mitochondria, but in the cytosol with the participation of the glyoxylate cycle. Now it is know that bacterial, plant and animal glyoxylate cycle is localized in small *glyoxysomes* [55, 56].

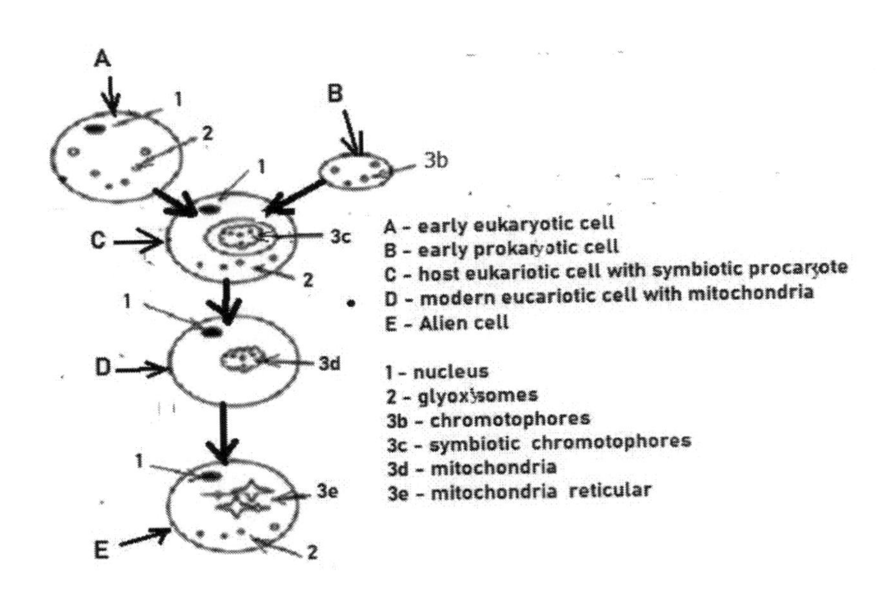

Mitochondrias evolution of eukaryotes and their transformation in Alien celles

How could such serious changes in Aliens cellular metabolism have occurred in course of evolution? The simplest eukaryotes did not have mitochondrias; their sugars were oxydized in cytosol glyoxysomes. According to theory of symbiogenesis, mitochondria in eukaryotic cells appeared as a result of phagocytosis of bacteria that had more productive oxidative systems in form of the Krebs cycle. Gradually, bacterial cells passed successively to symbiosis and parasitism, until they completely lost their independence, retaining, however, an autonomous system of protein biosynthesis. All these transformations took place at dawn of formation of a living world, and most of unicellular and multicellular animals, including termites, Alien ancestors, have full-fledged mitochondrias. However, Aliens mitochondrias had been undergone some metamorphosis. They have undergone reduction with loss of some functions, Krebs cycle was lost, although electron transport chains were kept only. Partially reduced small mitochondria were fixed on the surface of endoplasmic reticulum membranes. Perhaps this happened by heat or radiation stress. May be mitochondria structure and functions reduction have led to emergence of compensatory mechanisms in form of «forgotten» enzyme systems (dormant genes) of the ancient glyoxylate cycle.

So amazing information about altered Alien's mitochondrias is in full agreement with the data on the presence of a cytoplasmic glyoxylate cycle in their cells and a nontrivial macroerg GTP (instead of ATP). All this also testifies in favor of the reality of practical research of Aliens, reflected in the REPORT-96.

THERMOGENESIS

Unfortunately, how the Alien regulates body temperature is no clear. The heart rate of Aliens can be regulated by the splanchnic nerve, as well as by humoral factors. Mixed, venous and arterial blood circulates in a circulatory complex of Aliens, however, the degree of its mixing regulated by healthy smooth muscles under important nerve and humoral impulses, which increases susceptibility of a organism. Therefore, Aliens cannot be attributed to cold-blooded animals.

Parapharyngeal bodies secrete the active amine thermogranin, which stimulates heat production of fat pad cells when the ambient temperature drops. Mechanism of reactions is not described. In addition, thermogranin is one of blood clotting factors.

But there are a lot of mysteries in Aliens thermoregulation. Their eyes glow in the dark, and it is unlikely that this light was reflected. Witnesses argued that «something flashing» in their eyes. Paws of Aliens are hot and can even cause a burn. Snow in their traces usually was melted under their feet. Moreover, theses creatures, in particular, a Jersey Devil and a Jumping Jack, could douse the victim with blue «burning gas», may be releasing methabolic methane and hydrogen sulfide. Apparently, gases could be fired by a spark, like stingray electric discharge [5]. Perhaps Alien's unusual electrical characteristics are related to their cell membrane structure. Transmembrane transport of substrates cause disturbances of electromagnetic field of the membrane occur, leading to appearance of pores in the shielding layer of sulfo-lipid globules. Thus to study and evaluate materials of REPORT-96, participation of specialists from various biological specialties espeshelly biophysists is required.

NERVOUS SYSTEM

Alien's neurophysiology is especially interesting. The Alien's *central nervous system (*CNS), like all insects, is a chain of paired ganglia. The main differences between a central nervous system of Aliens and insects are in development and relative sizes of upper parts of primary brain, in presence of magnetosensor organs, as well as in structure of lower parts of CNS.

Despite real similarities to insects, it is obvious that at first, entomologists were hardly allowed to study Aliens. From the terminology used, it is clear that inicial nervous system investigation led by researchers of human Central Nervous System (may be of Pavlov Institute secret group, see below).

The most important part of Alien CNS is a forebrain (2). These are paired formations resembling a walnut with convolutions. The size of this part of the brain depends on the caste (in Upper castes it is 25 – 30 cm, in Workers in is 15 – 20 cm). Thus, a size of Alien forebrain significantly exceeds a human cerebral hemispheres. On section, the forebrain consisted of a layer of gray matter, beneath which were layers of conductive fibers. The forebrain is an associative center responsible for organizing the most complex forms of adaptive behavior, incl. speech and intellectual activity. Obviously, this part of Alien brain arose as a result of the phylogenetic development of insect mushroom bodies (1). Mushroom bodies of termites are responsible for formation of conditioned reflexes. The better they are developed, the higher the insect's ability to learn. After their destruction, insect loses all skills acquired during its life, although its unconditioned reflexes are preserved. The sizes of mushroom-shaped social insects also depend on caste [51].

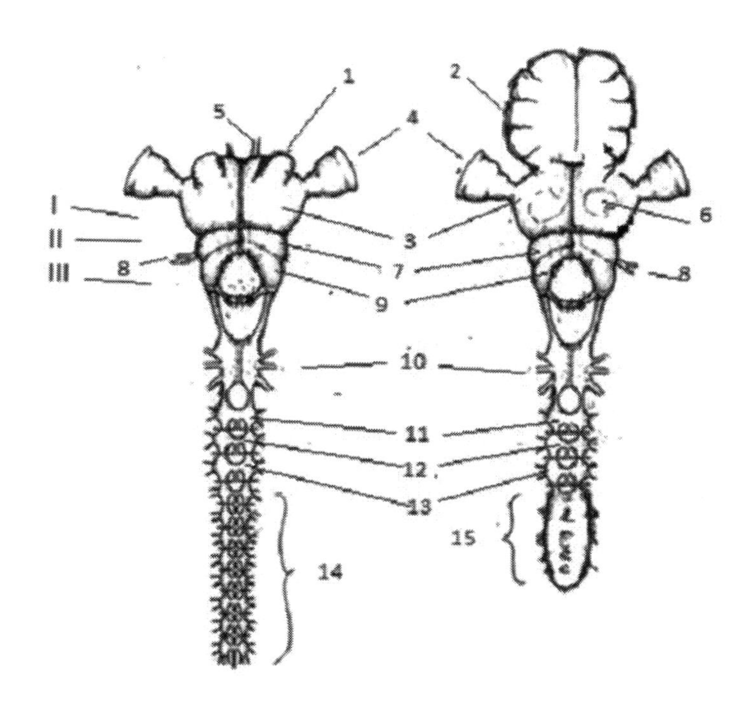

Insect and Alien Central Nervous System
(explanations in the text)

The middle brain of Aliens is a central plate, oliencephalon brain (3), connected to the paired lobes of a sensory brain. It contains paired optic lobes (4) and outgrowths innervating antennae (5, not shown in Aliens), antennules, and ocelli (not shown). The corresponding sections of Insect brains are arranged in exactly the same way, although they have different names. Unlike insects, a cavity filled with magnetolymph and nuclei of a magnetosensor organ (6). Organs of magnetic sensitivity have not yet been identified in insects, although it is known that magnetic sensitivity is present [80]. In a superpharyngeal region of the Alien brain there are centers innervating the antennae (8). The same centers in the corresponding area are also found in termites (7).

The somatic brain is located in a subpharyngeal region of Aliens. Here are the centers innervating an upper lip (9), grasping tentacles of a oral apparatus and paired lung sacs (10). A thickening of a somatic brain is below. A first termite thoracic ganglion innervates forelimbs. But Alien first thoracic ganglion innervates pseudocyrix (11). Even lower are pedal thickenings corresponding to a second and third thoracic ganglia, that innervate middle and hind limbs of termites and upper and lower limbs of Aliens, respectively (12, 13).

The lowest part of the Alien CNS is the visceral brain (15). This spherical formation corresponds to the six abdominal ganglia of termites (14), although it differs from them in shape. Here are the centers innervating internal organs of a viscerotheca (in insects – the abdomen).

Physiological characteristics of nervous systems of Alien and termites are similar. Excitement coming from several presynaptic fibers is combined, and summing up, overcomes a synaptic barrier. As a result action potential is generated in the postsynaptic axon in a certain rhythm. There is also ability to further block the synapse in response to monotonous stimuli or to facilitate transfer of excitation across a synapse. Aliens have a similar system of summation of a group of single charges. The subthreshold values of action potential can also be changed due to the additional opening of potassium channels. In addition, under influence of neurosecretion, a phase shift of action potential can occur, leading to excitation or inhibition.

These studies, especially the study of higher nervous activity of Aliens, had to be carried out in their lifetime. But Aliens were highly organized and very dangerous creatures. Behind descriptions of anatomical and physiological features of the CNS of Aliens, colossal and unsafe work of a large group of neurophysiologists can guessed.

EVOLUTIONARY ORIGINE of ALIENS

Evolutionary branch of insects appeared as a result of development of land by arthropods, this happened in the middle of the Paleozoic era, in the late Silurian or early Devonian, about 410 million years ago.

Virtual model of a last common ancestor of Polyneoptera

Polyneopters (Polyneoptera) are a large group of winged insects that lived about 350 million years ago. Grasshoppers, praying mantises, cockroaches, termites, stick insects, stoneflies and many other lesser known insects had a common ancestor. Now, phylogenetic reconstructions are increasingly being built taking into account the primary sequences of monomers in DNA or proteins. This makes it possible to determine much more accurately not only the degree of relationship of certain species, but also the time of their divergence. These methods are based on frequency number mutagenesis calculation. In general paleontological and molecular genetic data are in good agreement and complement each other. Recently, in a collaborative effort, entomologists, paleontologists, and molecular biologists have reconstructed an appearance of insects from the ancient group of polyneopters [58].

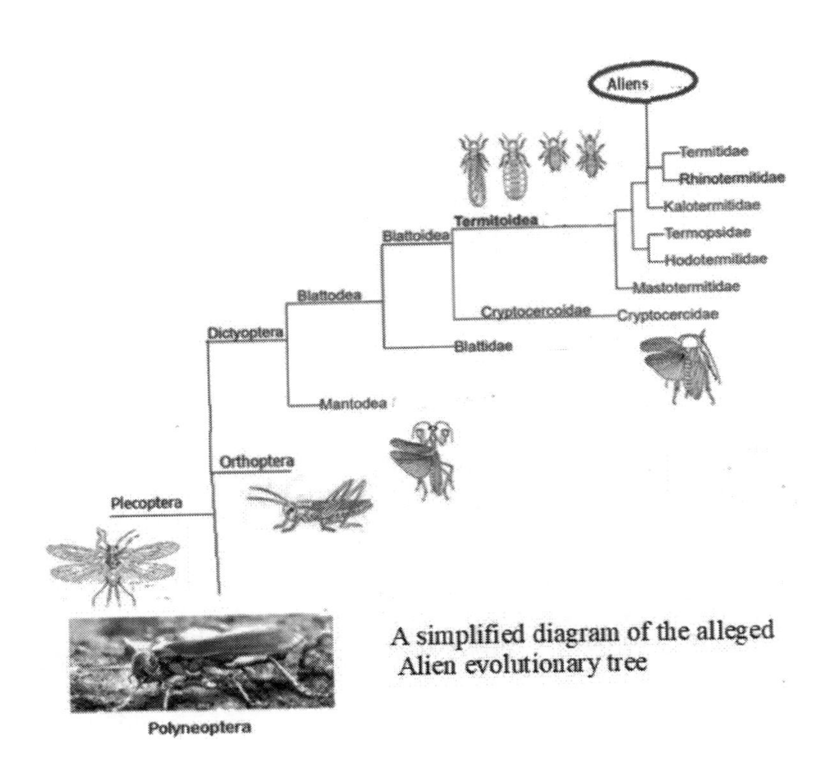

A simplified diagram of the alleged Alien evolutionary tree

The first representatives of this group had a gnawing mouth apparatus, a non-flattened soft body, and five-segmented legs. Their double wings were functionals, the upper ones had not yet been transformed into elytra. There were segmented appendages at end of an abdomen. Most likely, these insects were omnivorous and lived on ground. The growth and development of these primitive insects proceeded according to a simplified scheme, with «incomplete transformation». A more complex metamorphosis, including larval and pupal stages, appeared much later and was associated with difficulties of wing formation.

Polyneopters gave rise to many orders of insects, incl. Cockroaches (*Blattodea*) and Termites (*Isoptera*). The ancient cockroaches (*Dictyoptera*) were one of the first to bud off from this branch of the phylogenetic tree, and about 300 million years ago they gave rise to Mantis (*Mantodea*), and a little later – to Cockroaches (*Blattodea*) and Termites (*Isoptera*) and their more distant relatives, Orthoptera, which include grasshoppers and locusts, and stoneflies (*Plecoptera*). Methods of molecular genetics have shown a last common ancestors of cockroaches and termites existed 275 million years ago [59]. Zoologists still argue about order of these divergences. The figure shows only those groups of insects with which the Aliens have a similarity. The same diagram shows the alleged phylogenetic lineage of Aliens.

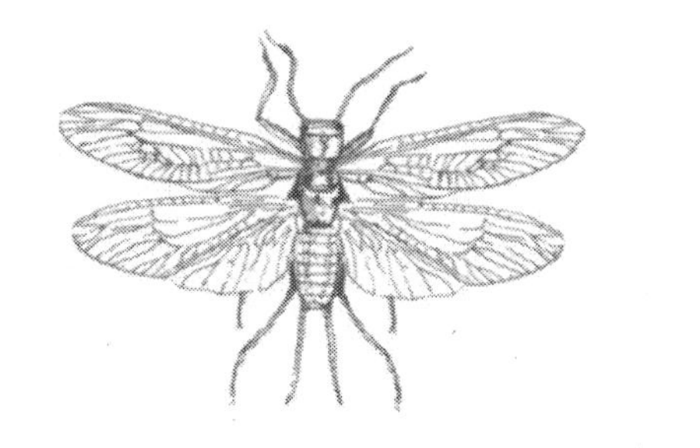

Stoneflies (*Plecoptera*) (left)
and grasshoppers (*Orthoptera*) (right)

Stoneflies (*Plecoptera*) are distant relatives of termites, but they have some similarities with Aliens. Nevertheless, Aliens are also united with them by presence of gills of larvae and hemocyanin in blood (in stoneflies – in hemolymph). However, larval gills in different species of stoneflies can be located both on the thoracic and on the abdominal segments of the body. And on a prothorax, like in Aliens, they are found only in genus *Jolia*. You could also pay attention to wings. In stoneflies, they are able to wrap themselves around a body («like a cloak», which was observed in Chupacabrus and Jumping Jack. See below). Therefore, hypothesis of Aliens origin from Stoneflies, apparently, should not be completely discarded. So far, there are only two arguments against this assumption: representatives of this order do not have any forms of sociality, and so far there are no examples of characteristic fragmentation of chromosomes.

On the other hand, Polyneopterans, the forerunners of all the above groups of insects, may have inherited gills and hemocyanin directly from crustaceans. In the process of further evolution, most insects lost these characters, at least in adult stage, imago. Nevertheless, germs of larval gills have been preserved even in termites. They were noticed by Fritz Müller while observing the development of South American termites (1875) [7].

Thanks to jumps and exterior, Chupacabrus were compared with grasshoppers. However, the Orthoptera group, which includes jumping insects with incomplete transformation, including grasshoppers, crickets, fillies, quails and jumpers, is closely related to Termites.

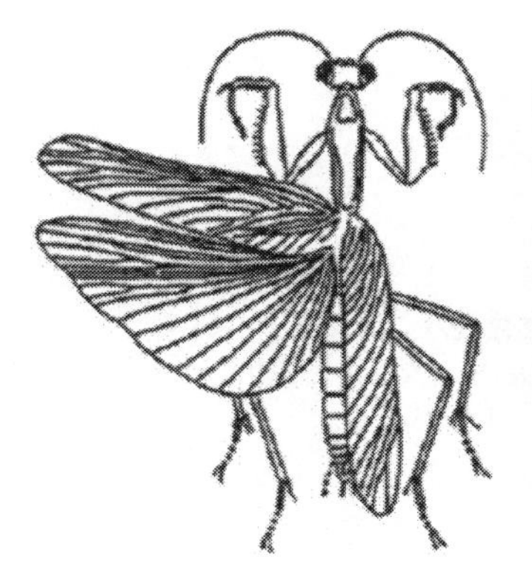

Mantis (*Mantodea*)

Ancient cockroaches (*Dictyoptera*) branched off from the common phylogenetic tree of Polyneoptera about 300 million years ago, they gave rise to Praying Mantises (*Mantodea*), and a little later – Cockroaches (*Blattodea*) and Termites (*Isoptera*).

Although Praying Mantises (*Mantodea*) and Termites (*Isoptera*) are highly specialized cockroaches that originally ingested organic debris and humus, mantises diverged earlier and were gradually turned to predation. Although Aliens are also predators, and this would seem to make them related, but any sociality of mantises is unknown.

Cryptocercus garciai (A) and
worker of *Mastotermes darwiniensis* (B)

The most important property of the cryptids described in REPORT-96 is their sociality. In evolution, this phenomenon arosed independently several times, and in animals groups which were systematically distant to each other. It is believed that sociality arosed in some cockroaches, before their isolation of a branch leading to termites. Sociality originated as an adaptation for distribution of symbiotic wooddestroying microflora among nest members. For example, females of a relic cockroach Cryptocercus, which have survived to this day, demonstrate complex social behavior [61 – 64].

However, Aliens are close just to termites, as they have not only in sociality, but also a similarity of colonies composition, of external and internal structures, as well as in characteristic fragmentation of chromosomes.

A primitive family *Mastotermitidae* also have a well-developed caste system, although along with it there aree also cockroach-like features [65, 66]. Their fossil wings have been found in Permian sediments in Kansas, other their remains were found in South America, Europe and Russia. Some specimens were very large insects, with a wingspan of about 5 cm. A relic Australian termite *Mastotermes darwiniensis* has survived till now. These termites are 10-15 mm. They are recognized as the most primitive species of modern termites [67, 68]. Apparently, their divergence occurred a very long time ago, because, unlike other termites, they have 96 chromosomes (other termites have less, 46 – 56 chromosomes) [69]. F.M. Wisner showed that the family *Mastotermitidae* appeared in the late Permian, more than 250 million years ago [65].

However, origin of Aliens from more advanced termite species seems more likely. Ancestors of Aliens have many similarities to the highly organized termite families *Rhinotermitidae* and *Termitidae*. For example termites of *Rhinotermitidae* have haploid workers, a factor similar to Alien's «factor of allocide-dependent transduction» and characteristic fragmentation of chromosomes. Some representatives of another highly organized family *Termitidae* have elso some features characteristic of Aliens. For example, Alien Small Worker mandibles are similar to that of the *Cubitermes* soldier. A «lophophore» is like to an elongated nose of the Rhynchotermes soldier. Termites of these families cultivate mushrooms in their nests.

Although these highly specialized families of termites were only widespread in the Jurassic, these features may have appeared in their earlier ancestors. In addition, according to some data, the emergence of highly organized families occurred more than 110 million years ago [62]. Although it seems that this evolutionary branch should have branch off even earlier (about 150 million years ago). May be a divergence time of Alien ancestors could be based on climatic conditions. Robert Berner of Yale University and Harry Landis of the American Geological Society calculated that atmosphere oxygen levels reached 35% by the end of Carboniferous. That is why giants existed among arthropods, their fossilized remains were found in England, France, Russia and North America. For example each leg of spiders Megarachne servinei reached half a meter, wingspans of the French dragonfly Meganeura monyi and the North American Meganeuropsis permiana had a of more than 70 centimeters.

In an environment rich in oxygen, insects living on land, as unnecessary, gradually reduced the gills. Larval gills are preserved mainly in primitive insects living in water. In conditions of excess oxygen, insect circulatory system have been significantly reduced. The oxygen-carrying protein hemocyanin, which was once present in hemolymph of their crustacean ancestors, lost its function, turning into a storage protein, hexamerin. Hemocyanin retained only in some species of larvae living in water. Such simplified system of respiration and blood circulation could be ensured existence of large insects in an oxygen-rich atmosphere only.

However, with the beginning of Perm, everything changed. At this time, there was a shift in the continents with significant rearrangements of landscapes and with inevitable climate changes. By the beginning of the Permian period, the oxygen level dropped significantly. A respiratory tracheal system of insects under such conditions no longer provided effective gas exchange. When oxygen content dropped at several times (from 35% to 7%), a great Permian extinction began. Many species of marine and terrestrial fauna could not survive it. This catastrophe became the only known mass extinction of insects, when almost 90% of species disappeared [70, 71, 83].

Those species of insects that nevertheless managed to survive have significantly decreased in size. That is why most modern insects, unlike other classes of animals, are less than a centimeter in size. However, some species of insects have undergone metamorphoses, which allowed them not only to survive, but also to develop successfully. Apparently, a volcanic activity of the Permian period, sulfurous gases, and abundant acrid rain drove many animals underground. Apparently, some species of ancient cockroaches or termites were even able to master underground isolates, where oxygen content was even lower. Some species preserved rudiments of gills and ability to synthesize hemocyanine. Existence in deep underground cavities with elevated temperature and radiation promoted mutagenesis and accelerated metamorphosis.

The tracheal «lungs» can be attributed to the important acquisitions of Aliens. The most important step in the evolution of Aliens took place when their ancestors acquired a closed circulatory system. The presence of hemocyanin, a specific oxygen carrier, contributes to the effective absorption of oxygen. All these changes allowed the Alien ancestors not only to survive in a low-oxygen environment, but also to noticeably increase in size.

Significant increase of oxygen mass occurred in late Triassic-early Jurassic. This mass was decreased from middle of Cretaceous period, however, it remained some high level until modern era. So, late Triassic can be considered as a second stage of Alien evolution. During this period, an increase in the size of the alien ancestors could have occurred due to new increase in the oxygen content in the atmosphere. For this group of insects, which have an effective respiratory and circulatory systems, their size has ceased to be a limiting factor.

Size increase could be of great importance not only for the survival of these species, but also for emergence of intelligence. Undoubtedly, the main prerequisite for its development was the sociality of Alien's termite-like ancestors. It is know, that populations of social insects such as termites, ants and bees have so-called «communal mind». Howewer a large brain masse is necessary for development of individual consciousness. Probably not the last role in development of Alien's nervous system was played by transition to predation and consumption of protein rich food.

In addition to sounds and smells, very important for Alien's communications, was magnetic sensivity and exchange of multidimensional images. A feature of this «vision» is that Alien sees not a flat picture, but perceives an object inside and out like a tomograph. Great changes passed in external structure of this group of animals. Alien's ancestors, after reduction of front forelimbs and their transformation into air sacs, had only two pairs of limbs. However, reason development and manipulation of objects upper limbs were necessary. So Alien's ancestors must be to move to upright posture. An internal skeleton (a chordal ring) appeared due necessity to support internal organs and body shape of such large animals.

Perhaps this intelligent species was formed in such evolutionary way. Evolution, which began about 150-200 million years ago, could lead to formation of a reasonable society and a peculiar culture by the middle of Jura, about 60 million years ago. Although Aliens are descended from insects, they are no longer such, and perhaps they should be attributed to a separate class of animals. Indeed, in its formation there were several aromorphoses. This predatory species unknown to our official academic science has successfully survived to this day. It should be called Giant Eusocial Insectoids (GEI). We believe that they caused cases of exsanguination of small animals that took place around the world in the late 20th – early 21st centuries [5].

AUTHORS and CONTRACTORS

There is no doubt that REPORT-96 was prepared by highly qualified Russian-speaking specialists. It is unlikely that the creation of a pseudo-scientific compilation on the basis of already known facts would require a serious financial investment (the maximum is an additional fee for several specialists and the cost of «office supplies»). In the former USSR, there were many scientific institutions of a suitable profile: national and republican Institutes of zoology, medecin and agricultural sciences and numerous branch laboratories. However, researches of Aliens required not only specialists, devices, and conditions for storing materials, but a high level of research secrecy. All these conditions were at Sechenov Institute of Evolutionary Physiology and Biochemistry of USSR Academy of Sciences (now Russian Academy of Sciences in St. Petersburg). It was a scientific institution with strict secrecy. There are a number of indirect evidences that just Sechenov Institute employees took part in practical study of Alien's biomaterials.

This Institute was created in 1956 on basis of a small laboratory of evolutionary physiology, under leadership of academician L.A. Orbeli (1882 – 1958). Its Studies were taking a look at development of different functions of animal organisms in onto- and phylogenesis. At first, the institute had only 9 laboratories.

Sechenov Institut staff: Anna K. Voskresenska
and Vladivir L. Sviderski

An unknown event that marked beginning of Alien research took place in 1965. Two years later, the Institute built a new building. The same year for some purposes a barometric chamber was equipped on on the territory of the Institute. And a year later, a technical block with an underground (!) bunker was built [72]. Very generous funding must have been obtained for construction and purchase of equipment.

Specialists were also required to study a unusual object. Several divisions of Sechenov Institute have previously dealt with certain aspects of invertebrate physiology, including a laboratory that studied physiology of nervous system, under leadership of Dr. Anna Kapitonovna Voskresenskaya. Nevertheless, to carry out research of this level sentient beings specialists in the field of higher nervous activity of human were required. Obviously, some rearrangements were urgently needed. Therefore, in 1966, a group of relevant specialists was transferred to Sechenov Institute from the Pavlov Institute of Human and Animal Physiology. Apparently, at first just this group was involved in study of Aliens (this is evidenced by using some non-traditional terms for invertebrate specialists).

Erelong, in 1967, Dr. Voskresenska died tragically in a car accident [73]. Soon part of her laboratory was merged with this group of specialists who arrived from the Pavlov Institute. A former employee of Voskresenska, Vladimir Leonidovich Svidersky was headed this secret department. In the past Svidersky, he was graduated with honors from Kirov Military Medical Academy, made a military career in Sevastopol and suddenly transferred to the Leningrad Zoological Research Institute. Open sources of Sechenov Institute state informed that Svidersky worked there since 1958 as a laboratory assistant (after our publications, this information disappeared from the official website of Sechenov Institute). However, this is unlikely, because a position of a laboratory assistant was always low-paid and unprestigious. May be former military was specially appointed and by 1965, even during Voskresenskaya's lifetime, Svidersky was already a young, promising PhD in biological sciences. In addition, Svidersky was a communist party leader of Sechenov Institute, an editor-in-chief of «Journal of Evolutionary Biochemistry and Physiology», founded in the same 1965. In 1981 Svidersky became leader of Sechenov Institute. Usually this job position is posseded the most relevant researcher, he receive the most funding. Svidersky subsequently made a brilliant scientific career, rising to Academician rank in Scientific Academy Department of Physiology. It is quite obvious that all research of the Institute were supervised by USSR security agencies.

Thanks to new «topicalities» the Sechenov Institute has grown, new employe positions and laboratories have been added.

Instead of 9 original laboratories by 2000, it included about thirty. Some unclassified aspects were published in open press [74, 75]. Later staffs of the Sechenov Institute used old resultes for a long time. With such closeness and security, the rank-and-file employees could not even guess about a true objects of research carried out at the Institute, only a immediate responsible executors could know about. However, it is unlikely that such detailed studies were so relevant for method development for combating agricultural pests. It is not clear, for example, how it is possible to substantiate from this point of view the effect of the gravity on locomotor centers of agricultural pests («Changes in the activity of locust Migratoria locust locomotive centers under the influence of increased gravity») [76]. Perhaps the task was to find out Alien's adaptation ability to space flight?

From the above, it is clear that the Sechenov Institute at the turn of the 70s received serious financing. State awards also indirectly testify to significance of Svidersky Department research: in 1987, he was awarded the State Prize and the Order of Honor. Exactly then a report summarizing results of research must have been presented. So, the text of REPORT-96 must be prepared just in 1987.

COINCIDENCE OF THE RESULTS OF SOVIET AND AMERICAN STUDIES

Judging by some signs, American researchers also worked just with termitoids. Former president of the Washington Institute of Technology Dr. R. Sarsbacher, was a consultant to the Pentagon in the middle of last century. In the service, he had to communicate with some employees of Vannevar Bush, Robert Oppenheimer and John von Neumann (members of a secret group, hypothetical so-called «MJ-12»). In private communication, he received information about the existence of insect-like aliens. Their bodies were very light but strong, their skin hard. Under microscope, it looked like a «wicker» and probably represented chitinous covers of different thicknesses. Dr. Sarsbacher reported this in a letter to American ufologist William Steinman dated November 29, [30].

American researchers been aware of Alien nature:
Dr. D. Burish (left) and Dr. R. Sorbacher (right)

The above-mentioned report by American researchers Q94-109 is just one of several documents containing fragmentary information about these studies. Later, their author, Dr. Danny Benjamin Crain, Ph.D., released many interviews where he gave confusing explanations about histological and biochemical features of Alien tissues [77]. Now, after release of sensational materials, this doctor got better known as Dan Burish (Dan B Catselas Burisch, PhD). Research was allegedly made in deep underground laboratories of US Air Force military bases. The document states that this work was led by Dr. John Anthony McGuinness, a customer of research was Rear Admiral J. McConnell (one of the members of MJ-12 hypothetical group), and the research itself was carried out as «Aquarius project» part.

According to the initial information, the direct executors did not know about the true nature of the object of study [78]. A completely different secret group was preparing samples for research. However, it is obvious that tissue samples were similar to that of the Sechenov Institute. Alien cells were very large, multinucleated perikaryons and were connected to each other by processes that form an «outer cytoplasm». Channels passed through the processes, which stretch through the entire cellular matrix. Report Q94-109, like REPORT-96, noted that granules, «phylomenas», were present in the submembrane layer of the Alien cell cytoplasm and contain spare material for the restoration of the plasmalemma. Sechenov Institute and American researchers believed this material was depleted with age. The Americans even tried to cure this «age pathology» by using various fusion methods of Alien and human cells (which was a futile undertaking).

Given the many overlaps, did this information have a common source, or did the Soviet and American teams work independently? The data of the Americans organically complement the work of researchers from Sechenov Institute. However, the report of Russian experts is worked out much more fully and is logically consistent. For example, the partial reduction of mitochondria in Aliens is in good agreement with the presence of the glyoxylate cycle in their metabolism. This cycle is analogous to the Krebs cycle (which in eukaryotic cells occurs in mitochondria during the metabolism of sugars), but is localized in cytosol glyoxysomes. Based on the information presented in REPORT-96, it is possible to construct a hypothesis how metabolic changes could lead to an increase in the size, structure modification and cephalization of a new arthropod species. May be the fact of the high role of GTP in Alien cells was discovered by Russian and American researchers independently, but American researchers explained it on base datas, obnained in 1994 [82].

It is possible that American researchers were working with specimens captured at UFO crash sites, for example, in January 1996, in the Brazilian Varginha [81]. It is possible that the American cytologist Dr. Burish worked with such material.

However, it is more likely that Aliens were captured somewhere in underground cavities, where their nest-cities, technological premises and other infrastructure are hidden.

Studies of the processes occurring in the central nervous system, and, in particular, mechanisms of memory storage, could be carried out in vivo only. And this means that whole groups of representatives of different castes, are captured somewhere and somewhere are kept. Most likely, the alien population was discovered during the sinking of mines or when blasting was used in the development of deposits. We know from REPORT-96 that Insectoids were discovered in 1965. According to one version, this happened after explosions during the drifting of one of the mines, according to another source, as a result of a specially organized capture operation. The exact location of an incident is still unknown, however, there is reason to believe that it happened on the Kola Peninsula or in Central Asia (according to information by N. Subbotin).

NOTES to PART 1.

1. FidoNet or Fido for short is an international amateur non-commercial computer network, built using point-to-point technology.
2. «Anatomy and Physiology of Aliens», Internet-edition «X-libri of Major Thomas», www.x-libri.ru/elib/innet237/index.htm
3. subscribe.ru/archive/rest.mystery.secrets/200011/16020601.html
4. Unknown Chupacabrus vampires, Jersey Devil and Moth-Man predators, apparently, are the Small Alien Workers, representatives of the foraging caste who go hunting at night. In any case, jumping, reaction to fright correspond exactly to how Chupacabrus behave [5].
5. Agon E, Shevik S, Alkor E – «The Mysterious Devil Chupacabrus», Toronto, Altaspera, 2020, 295 pp. ill.
6. Aanen DK, Eggleton P, Rouland-Lefevre C et al. // PNAS, 2002, 99, 14887-14992
7. Mueller UG, Schultz TR, Aanen, DK et al. // Ann. Rev. Ecol. Evol. Syst., 2005, 36, 563-595
8. In Hymenoptera, bees, ants, females, queen and workers are diploid, and males, drones, are haploid. The differences

between the reproductive female and the workers only appear because of the way they are fed.

9. Rodendorf B, Rasnitsyn A – «Historical development of the class of insects» // Proceedings of the Paleontological Institute of the USSR Academy of Sciences, vol. 175. M, Nauka, 1980, 269 p.

10. Yoshinobu Hayashi, Nathan Lo, Hitoshi Miyata, Osamu Kitade // Science. 2007, 318, 985-987

11. Vincke PP, Tilquin JP // Chromosoma. 1978, 67, 151-156

12. DeHeer CJ, Vargo EL // Insectes Soc. 2008, 55, 190-199

13. Zhuzhikov DP – «Termites of USSR», M, Publishing Moskow University, 1979, 225 p. ill.

14. Perhaps the names were chosen for secrecy motives when to give tissue samples to laboratory assistant, or texts on tiping.

15. On the head of insectoid from the photo, there are also some processes. However, they are not similar to what is described in REPORT-96. They look like paws. (There is a suspicion that these paws are false). The other hand, a mutation in Antennapedia gene may cause a abnormal arrangement of limbs (such mutilations led to the formation of limbs in place of the antennae in Drosophila melanogaster) – Pierce BA – «Genetics: A Conceptual approach», 2nd edition, 2004

16. Antennas, or «small antennas». In the case when an arthropod has large and small antennae, the smaller ones are called antennae.

17. It is possible that these formations resembled «ridges» to some abductants [18, 19, 34]

18. Turner K «Into the Fringe», 1992

19. Turner K, Rice T «Masquerade of Angels», 1994

20. Rocha MM, Morales-Corrêa-Castro AC, Cuezzo C et al. // PLoS One. 2017, 12, 3, e017436

21. Although there is no clarification in REPORT-96, it is known from other sources that the mouth organs and eyes of the representatives of Upper castes are different. Their mouth opening is a slit, and their faceted eyes are much larger. Soldiers can not have eyes or have small ocelli.

22. Chupacabrus pierces the skin and muscles of the victim with its long mandibles, reaching large vessels, where it injects

hydrolytic enzymes. After a few minutes, he aspirates the enzyme-treated blood [5].

23. An eyewitness in Puerto Rico saw that the Chupacabrus «swelled his goiter» before takeoff [5].

24. Zakhvatkin YA «Course of General Entomology», 2001.

25. In fact, F. Müller's hypothesis complements the Oken-Gegenbaur hypothesis that the tracheal gills are the predecessors of the wings. Müller also suggested that the outgrowths of the posterior and mesothorax turn into wings (1875).

26. Müller F., Haeckel EG «Basic biogenetic law», 2013

27. Similar formations are also described by some witnesses; They seemed to David Turner to be something «like scraps of skin or a scarf wound around the head» [18].

28. Pseudoxyryx is a tube with a blind closed end.

29. Sounds made by Aliens resemble gnashing [30], buzzing or «booming» [31]. the monsters in Varginha also made a buzzing sound.

30. Howe L «An Alien Harvest. Further Evidence Linking Animal Mutilations and Human Abductions to Alien Life Forms», 1989

31 Obiedkov O «UFOs in the Orenburg region: UFOs over Orsk», 1990

32. Eskov AK // Soros educational journal, 1997, 11, 22 – 29

33. It is known from other sources that the upper limbs of Aliens are able to bend in a non-trivial way in the form of the Latin letter «S», in any case, some abductants spoke about this [34]. Perhaps this is a consequence of mutations in homeotic genes [15].

34. Turner K «Taken: Inside the Alien-Human Agenda», 1994

35. Such features of the foot were attributed to the Owl-Man, Jersey Devil, Chupacabrus. Walking on such limbs makes a strange impression. Aliens, according to eyewitnesses, walk like a soldier on a march, raising their legs high [5].

36. It is known that some Chupacabrus (Chupacabra, Jersey Devil) not only had wings, but also skillfully used them. These wings were on the back and completely separate from the forelimbs. At least three witnesses from different parts of the world said that they are double, like a dragonfly or a moth [5].

37. Chitinases usually have a broad pH optimum [38].
38. Zhuravleva NV, Lukyanov PA // Vestnik FEB RAS. 2004, 3, 76 – 86
39. Perhaps this explains the terrifying smell of mercaptans and ammonia that Chupacabrus exudes, noted by many witnesses [5].
40. Bey-Bienko GI «General entomology» // M, Prospectus of Science, 2008.486 p. , ill.
41. Tyschenko VP «Physiology of insects» // M, Higher school, 1986. 303 p., Ill.
42.www.medvuz.com/noz/274.php
43 Burmester T, // Micron, 2004, 35, P. 121-122
44. Hagner-Holler S, Schoen A, Erker W et al. // PNAS, 2004, 101, 3, 871-874
45. Fochetti R Belardinelli M Guerra L et al. // The Protein Journal, 2006, 25
46. Burmester T, Scheller K // J. Mol. Evol. 1998, 47, 93-108
47. Kusche K // Eur. J. Biochem. 2003, 270, 2860-2868
48. Theopold U, Schmidt O // J. Insect Physiol. 1997,43,667-674
49. Scherfer C, Qazi MR, Takahashi K et al. // Dev Biol., 2006, 295, 1, 156-163
50. Natochin YV // Bulletin of the Russian Academy of Sciences, 2007, 77, M 11, 999-1010
51. Dogel VA «Comparative anatomy of invertebrates». Part 1. Leningrad, 1938 600 s, ill.
52. Nelson DL, Cox MM // «Principles of biochemistry». Fifth edition. New York: WH Freeman and company, 2008, 1158 p.
53. Burnell A.M., Houthoofd K., O'Hanlon K., Vanfleteren J.R. // Exp Gerontol. 2005, 40,11, P. 850-856
54. Kondrashov FA, Koonin EV, Morgunov IG et al. // Biol Direct. 2006, 1, P. 31
55. microbiology.ucoz.org/index/bakterii/0-11
56. Hermann HR (ed.). (1981). Social Insects. Volume 2. New York: Academic Press, 1981. XIII, P. 1-502;
57. Deligne J, Quennedey A, Blum MS « The Enemies ad Defense Mechanisms of Termites», P. 2-76

58. Wipfler B, Letsch H, Frandsen PV et al. // PNAS, 2019, 116, 8
59. Legendre F, Nel A, Svenson GJ et al // PLOS-ONE, 22 July 2015
60. Fontana F, Amorelli M // Bollettino di zoologia (Italian Journal of Zoology), 1975, 42, 99-100
61. Kenji Matsuura «Sexual and asexual reproduction in Termites» // Ed. David Edward Bignell, Yves Roisin, Nathan Lo. Biology of Termites: a Modern Synthesis – Springer Netherlands, 2011, 255- 277
62. Harrison MC, Jongepier E, Robertson HM et al. // Nat Ecol Evol. 2018 Mar; 2 (3), 557-566
63. Maekawa K, Nalepa CA // Insects. 2011 2 (3), 354-368
64. Nalera CA, Byers GW, Bandi C et al // Ann. Entomol. Soc. Am. 1997, 90, 4
65. Weesner FM (1960) // Annual Review of Entomology. 1960, 5, 1, 153-170
66. Legendre F, Whiting MF, Bordereau C et al. // Mol Phylogenet Evol. 2008 Aug; 48 (2), 615-627
67. Inward D, Beccaloni G, Eggleton P // Biol Lett. 2007, 22, 3, 3
68. Goodisman MAD, Crozier RH // Evolution. 2002, 56, 1
69. Bergamaschi S, Dawes-Gromadzki TZ, Scali V. et al. // Chromosome Research. 2007, 15, 6
70. Sole RV, Newman M – «Extinctions and Biodiversity in the Fossil Record», Volume Two, «The earth system: biological and ecological dimensions of global environment change», 2002
71. Benton MJ «When Life Nearly Died: The Greatest Mass Extinction of All Time», 2005
72. «Draft Program for the development of material and technical base Sechenov Institute of Evolutionary Physiology and Biochemistry of the Russian Academy of Sciences for 2013-2020».
73. Sveshnikov VG «Bylytsy», «Ridero», 2018, 1060 p.
74. Sviderskiy VL, Plotnikova SI // Zhurn. evol. biochem. and fiziol. 2002, 38, 5, 492-501
75. Gorelkin VS, Severina IU // Zhurn. evol. biochem. and fiziol. 2004, 40, 6, 508-516

76. Gorelkin VS, Severina IU, Kuznetsova TV // Zhurn. evol. biochem. and fiziol. 2001, 37, 4, 286-289

77. «The Q94-109A document authored by dr. Dan Burisch aka Dan Crain» – www.skywatch-research.org/Q94-109A.htm

78. According to initial information from Dr. Burish, the histologists did not know which object they were examining. The histological samples were prepared by a completely different team of researchers. Later, most likely under pressure from the secret services, Burish told a completely fantastic story, how he personally took tissues for analysis from a «humanoid» on one of the underground levels of the S-4 «Dreamland» base (also fake). He allegedly was admitted to such a dangerous procedure because he had «contacts» with the aliens for a long time. Everything that accompanied this story resembles an accidental «leak», which they later tried to cover up with misinformation. And although there are a number of questions to the project itself, and to some of Burish's statements, this information can be considered reliable in the part that concerns the morphology of Alien insectoid cells.

79. Feoktistova NY «Arms race among termites» // Biology, No. 36/1997

80. P. M. Jacklyn and U. Munro. Evidence for the use of magnetic cues in mound construction by the termite Amitermes meridionalis (Isoptera : Termitinae)//Australian Journal of Zoology 50(4) 357 – 368 Published: 15 November 2002

81. www.ufoevidence.org/cases/case100.htm

82. Randazzo PA, Kahn RA // J. Biol. Chem., 1994, 269,10758-10763

83. Goldblatt C, Lenton TM, Watson AJ «Bistability of atmospheric oxygen and the Great Oxidation» // Nature, 2006, v 443, P 683–686

PART 2. UNDERGROUND OPPOSITION

SEARCH FOR UNDERGROUND NESTS

Where are habitats of strange creatures described in REPORT-96? There is a direct reference to some underground «structures», which the authors of REPORT-96 call «nests». It should be noted that researchers became interested in strange underground activity before mankind wondered who owns «fires» and «saucers» flying above the earth with signs of intelligent behavior. Meanwhile, both are manifestations of a same really existing infernal world. The search for something unknown in caves, dungeons and old abandoned mines has been carried out since epy end of the 19th century. In Russia archaeologist and speleologist Ignatius Stelletsky (1878-1949) explored underground cavities, but his materials were confiscated by security agents Emergency Committe [1] of new communistic authorities. Later, they repeatedly attracted Stelletsky to work, witch was supervised directly.

In 1920-1922, an expedition of Bekhterev Brain Institute led by Alexander Barchenko (1881-1938) visited the Kola Peninsula. The purpose of the expedition was to study causes of a strange mental illness, which affected locals. This disease was called «meryachenie». Barchenko assumed that somewhere in Lake Seyd and the Lovozero tundra region there is a source of unknown electromagnetic radiation emanating from under the Khibiny mountain range. When exploring this area, the expedition discovered the exits of mysterious caves. After returning, the report of the expedition was immediately classified and Barchenko himself was invited to collaboration to People's Commissariat of Internal Affairs. The head of the Special

Department, which included Barchenko's laboratory, was Gleb Boky (1879-1938). Chekists tried to study nature and sources of geomagnetic anomalies distorting electromagnetic communication. This work was classified, and all employees were selected carefully. Barchenko and Bokiy organized several expeditions to the Kola and the Crimean Peninsulas, to the Volga Region and Altai. Chekists were interested elso in caves located under mountain ranges.

In 1937, Bokiy and Barchenko were arrested and soon shot. Was this haste due to their involvement arrested in the conspiracy, as the prosecution claimed, or was there an attempt to hide dangerous information? According to one version, this special unit, created in structure of security agencies, was able to operate and even conduct military operations underground. According to legend, this group suffered significant losses during some kind of special operation in caves of Central Asia in 1933 or 1934 [3]. Perhaps this execution was a consequence of these events? Be that as it may, this information was keeping in closed archives of Security Agencies of the USSR. In 1935 most of the known entrances to various dungeons and caves were blown up by military.

Even earlier, in November 1929, Yakov Blyumkin (1898-1929) was also hastily shot. Being an a senior authorized officer of the Foreign Department of Commissariat of Internal Affairs, he repeatedly visited countries of the Middle East, Tibet, Afghanistan and India on a secret mission. There he supposedly received intriguing information about some ancient underground inhabitants and «divine weapons» from Tibetan monks. Late Soviet investigation found that Blumkin shared secret materials with a representative of German military intelligence, receiving a very considerable amount of money.

Whether this is true or not, in the future Blyumkin's information and results of Barchenko's work interested of the German special services. Since the 1930s, German scientists and the military, led by Ernst Schaeffer, have been looking for some artifacts in Tibet. The Department of «Ahnenerbe» Institution, by leadership Dr. Hans Brand studied karst failures and underground cavities carried out on almost all continents. In search of non-traditional knowledge, from about 1938 the Germans sent their paramilitary groups to the Kola Peninsula and other mountainous regions. During the Second World War, German researchers worked in the Ukraine, the Crimea, and the Caucasus (1941 – 1943). The materials of their searches were still classified [2-5]. German military even reached the northern Urals. In any case, in the Sverdlovsk region, approximately where a dozen years later the tourist group of Igor Dyatlov will be died, a secret German airfield was discovered. After the Second World War, part of the German secrets migrated to intelligence officers of Western Coalition countries.

It is widely known that since 1947 the US military has been studying an UFO phenomenon. It is generally accepted that initially, study the phenomenon was carried out by the US Air Force intelligence. May be at first the intelligence officers from the Air Force and the NSA tried to establish contact, them selves poorly understanding what, in fact, they were dealing with. However, already in the early 1950s, it became clear that the aliens showed hostility, preying on animals and people. According to retired Air Force Lieutenant Colonel Philippe Corso, the US authorities have been aware of these terrible phenomena since 1951. In 1952, the CIA also joined the study of the problem [6]. In the same year, the US authorities, realizing the seriousness of the situation, considered it necessary to inform the leaders of the leading countries of the world, including the USSR, about this problem.

At the same time, American authorities tried to hide it from the population. According to the recommendations of the Robertson Commission, assembled at initiative of the CIA in 1953, any information about «aliens» should be hided and develop a frivolous attitude towards «extraterrestrials» to prevent possible panic among the population [6, 7]. By that time, despite the high secrecy, the projects concerning aliens were carried out very intense, and military attitude towards the problem was more than serious.

The military tracked UFO flights and even allegedly shot down some flying objects [8, 9, 10] (see PART 3). Despite the fact that in the United States Aliens were quick to declare arriving from distant stars, the surveys were carried out not in air space only, but military works were carried elso in underground.

In the midlle 1950s, tunnels and mines were being drilled in the United States, and underground multi-level complexes were being built. The construction of such multi-level underground facilities, apparently, had many purposes, although the projects only mentioned of using them as shelter in the event of a global world war. The initiator of research in the field of deep underground construction was the US Air Force. The construction was carried out as part of secret programs and cost the military budget more than 12 trillion dollars. The work was carried out by the RAND Corporation.

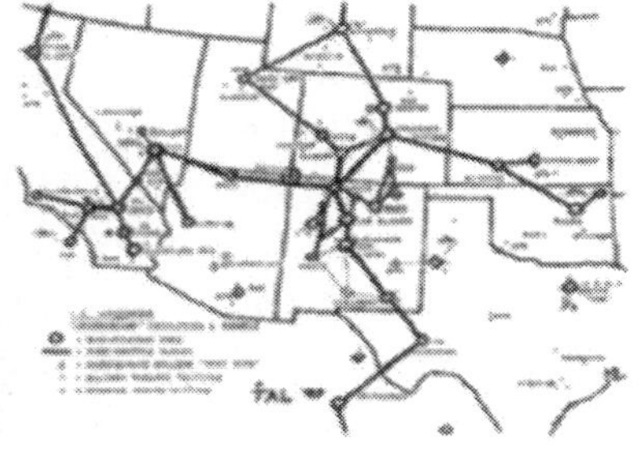

Underground drilling machine in Nevada (1982)
and map of underground military base locations
in South-Western States (by Valerian Tal, 1990)

Huge tunneling machines were involved in tunnel construction (there is evidence that similar equipment was also tested in the USSR [12]). Soon, multi-tiered underground complexes were built, for example, in Indian Springs in Nevada [11]. The attention of researchers involved in the problem of UFOs has long been riveted to such alleged bases, located near Groom Lake in Nevada and in the Dulce of New Mexico, etc. It is believed that downed UFOs and remains of their crews were researched there. An analysis of available information showed that the locations of military bases practically coincide with areas of UFO activity, with places of anomalous killings of livestock and disappearances of people. According to Andrew Thomas, this is no coincidence. In the same areas, gravitational anomalies are observed, indicating a possible location of underground voids [11]. Timothy Good, told BBC2 a strange story that President Dwight Eisenhower allegedly met with «extraterrestrials» at Edwards Air Force Base in California in 1954 and agreed to share the underground bases. Despite this story has been discredited, many ufologists continue to believe in it. However, in really why are bases built underground?

It is highly likely that *underground military bases are being built in close proximity to what military slang used to call «anthills»* (or «termite mounds»), i.e. places of underground nests of Aliens. That is why such «termite mounds» are mistakenly identified with military bases. Aliens are the indigenous inhabitants of the Earth, belonging to an ancient biological species, and the dungeons are its natural habitat. Therefore, in order to avoid further confusion, «bases» will be called exclusively military buildings, and the underground structures of aliens – «nests», although in some cases it is difficult to determine who still owns the underground structures.

Dr. Richard Sauder, a political scientist, a former employee of the Department of Natural Resources, has been searching for underground tunnels for more than 20 years. He stated that such objects really exist. They stretch for many kilometers underground, and, in his opinion, connect countries and continents. He believes that these underground communications lead to inhabited «cities». In the United States, special military teams have been created to search for such underground objects, mask entrances (portals) and inspect tunnels. If there is any activity in them, they are blocked. Military bases are set up near such portals. If the objects are uninhabited, they are used for military purposes [13]. Physicist Dr. Paul Bennewitz was also sure that near Dulce, there is a nest in which aliens are present [15]. In some cases, people trying to get to the nests ran into military guards. For example, this happened in 1988 with two digers trying to get into the nesting of Aliens under El Cahul mountain in Puerto Rico, making their way there through the ventilation shaft [11]. A man named Timothy, who in the 1970s served as a junior officer at one of the underground military bases in Cold Lake in Canada, was involved in an investigation related to the killing of cattle in Alberta. He was also acquainted with secret information, according to which the nests of predatory creatures that inhabited the Earth long before the appearance of man are scattered all over the world. About three dozen of them are in Canada, one hundred and forty – in the United States and more than one and a half thousand – in other countries [14].

Philipp Schneider, an American mining engineer of German origin, spread information about a secret treaty of American authorities and aliens about common bases construction. However, an analysis of his account of 1979 events suggests otherwise. In 1979, Schneider was involved in construction work to «expand» an already existing military base in New Mexico. It was strange that underground work was carried out not near a military base, but aside, in the desert. Surprising also was presence of a large number of green berets. Apparently, the base command was ready for any non-standard situation. Military builders carried out preparatory blasting. The danger arose when an explosion occurred that destroyed walls of an underground «termitary». Schneider, descending underground to inspect the rocks, stumbled upon its inhabitants. Thus began an incident in which about seventy people from special forces and base employees were killed. Schneider was severely injured, receiving very severe burns [16].

All this is not too reminiscent of peaceful cooperation, it is more likely that the military tried to penetrate where they were not allowed, namely, into underground cavities, nests of an alien civilization. So such a close neighborhood military and alien nesting sites may be a perfectly reasonable explanation. All activities of the military in this direction are top secret. Some ufologists believe that the cover-up is due not so much to the strategic tasks of the military as to preventing the publication of the fact that the military is not yet capable of resisting foreign forces. However, secrecy, designed to prevent panic among the population, has its downside, when crowds of militant idiots storm military bases, believing that it is there that they will find all the «alien» secrets.

There is reason to believe that the military of the USSR was also very seriously preparing for the confrontation with the aliens. According to a former military man who served in 1987 in the «Office of the Chief of Space Facilities» and who, for obvious reasons, wished to remain anonymous, around the end of the forties, the Soviet Union began to develop weapons that would make it possible to counter aggression or hostile actions «from space «. These projects, like the nuclear ones, were led by Lavrenty Beria, but the level of secrecy was an order of magnitude higher ... The project had two components: chemical-biological and physical. In the first direction, chemical agents were developed that are harmless to humans, but affect life forms based on a different metabolism. In the second direction, the atomic bomb was considered as a universal weapon capable of deterring any aggressor. However, later the direction of work was changed, and the project grew into a completely different area [17].

«Another area» was underfoot. The Soviet secret services were well aware of the existence of cyclopean underground structures. The old underground buildings abandoned by the aliens are successfully used by Russian troops as secret warehouses and for the covert movement of military supplies and units. Many secret underground abandoned paths exist in the Chelyabinsk region, in Bashkiria, in the north of the Krasnoyarsk Territory and in a number of other places in Russia.

However, not all underground structures are abandoned by Aliens. It seems no coincidence that in 1965 the USSR also adopted a program of underground nuclear explosions, which was in effect until 1988 [19]. Its implementation was carried out by specialists from two secret nuclear centers: «Arzamas-16» and «Chelyabinsk-70». Traces of atomic strikes and radiation have been preserved in bowels of many regions of Russia. For example, on September 19, 1971, an underground nuclear explosion was carried out at a depth of about 600 m, 350 kilometers from Moscow, near Shacha village of Ivanovo region. Interestingly, almost immediately, luminous UFOs appeared in the sky above this place. «Peaceful» underground nuclear explosions were also heard on the Kola Peninsula. The first nuclear charge was detonated in Khibiny in September 1972. It was made in the bowels of Mount Kuelporr, and this operation was code-named «Dnepr-1». In August 1984, the operation called «Dnepr-2» was continued. In the bowels of the same mountain, a second nuclear explosion was made, already by two nuclear charges. It was horizontal and directed towards Lovozero. Some researchers cite evidence from local residents that underground explosions were carried out both directly in Lovozero and in the bowels of highest mountain Angvundaschorr.

Although according to the official version, the purpose of these experiments was the crushing of apatite ore, this is unlikely. Then about 400 thousand tons of ore were beaten off. But necessary roads were not built for the export of ore mined. The high radioactivity of ore and waste did not allow it to be used in enrichment plants. Based on this, member of Environmental Safety Council prof. Alexei Vladimirovich Yablokov (1933 – 2017) believed that a main goal of experiments was completely different. One employees of Scientific Department of border troops Makhov A.V. have admitted that «... There was some kind of significant, state the scale of the problem is most likely a danger. And this danger was hidden in almost all underground part of the Lovozero tundra. To get close directly to its source was impossible – man was overwhelmed by animal fear... The problem could only be solved with directional longitudinal-wave nuclear explosion».

These goals were quite serious. Directed underground nuclear explosions in the Khibiny were carried out by decision of the top Soviet leadership in order to destroy the most powerful underground source of unknown energy, located under the Lovozero mountain tundra and posing a threat to people. Strange as it may seem, after directed nuclear explosions, the residents of Lovozero region stopped bouts of «measuring» (a disease, which was studied by of Alexander Barchenko expedition, 1920).

In addition to underground nuclear explosions, ultra-deep drilling was also carried out on the Kola Peninsula on top-secret object (1970 – 1992). In 1990, the maximum depth was reached – 12262 m. Two years later, drilling was stopped, and the well was mothballed. Once there was an explosion at the well, the cause of which was never established. Later Drilling Director academisien David Mironovich Huberman admitted: «Indeed, a very strange noise was recorded, suddently there was an explosion and a few days later, nothing was found at the same depth» [21].

Fantastic stories on the Kola Peninsula are reinforced by very strange incidents, when local soldiers observed an insectoid or tourists died for some unknown reason [22, 23]. It is possible the purpose of ultra-deep drilling, as well as underground nuclear explosions, in some cases could be to capture Aliens.

All peoples have tales about terrible creatures living in the darkness of the dungeons. Exploring dungeons has always been considered a dangerous activity. Residents of villages located near strange pits have always complained that livestock disappears there. Sometimes people disappear without a trace. Notoriety goes about the Kobyakovsky caves of the Rostov region. Appearances of hideous red-eyed monsters have been observed in the area. Some of the caves of the Kobyakovsky settlement are very long and branched and look more like artificial passages. A terrible story was told by the researcher of the Kobyakovsky caves Vyacheslav Borisovich Zaporozhtsev, explorer of caves, underground passages and other similar objects in the vicinity of the city of Aksai. Soldiers died in 1949, when the military was examining underground caves and passages at the Kobyakovsky settlement and Esina gully.

People have repeatedly disappeared in the Peruvian dungeons of Cuzco. It is believed that in the 20s of the twentieth century several expeditions disappeared there, incl. and a scientific expedition from the University of Lima. Few of those who returned spoke of ingenious traps and terrible monsters. Some have lost their minds. The attacks of underground monsters aimed at uninvited guests are described [20]. May be the stories by Ann Kohan, who allegedly visited a Glowing Skulls cave in Honduras, and by spouses Marshall, who went to Cazo Diablo near Bishop town in California, based on real events [24]. Dangers await people not only in the dungeons, but also near them. American conspiracy theorist and former police officer David Polides compiled a map of places of disappearance, the main places of disappearances coincide with the alleged underground cavities [18] (see Part 3). The authorities of the leading countries are aware of the serious danger emanating from underground and are trying to block access for the uninitiated. Sometimes even archaeologists are forbidden to work in such places. When rock collapsed at a coal mine near Heavener, Oklahoma, at a depth of about 3 km after a technological explosion, exposing a wall of polished blocks, the miners were immediately evacuated, and the shaft was covered with waste rock and walled up (1928) [20]. Jan Paenk, a Polish researcher, cited the memoirs of a miner, according to which, while driving drifts and laying explosive packages, their team stumbled upon two unknown tunnels. By order of the authorities, they were quickly cemented.

In 1935, many entrances to dungeons and caves on the territory of the USSR were blown up by People's Commissariat of Internal Affairs. Such walling up and backfilling of tunnels and dungeons occurred repeatedly in various places on the territory of the USSR. Apparently, such precautions were justified and dictated not only by security considerations for amateur speleologists.

Seriously about the underground buildings of an unknown intelligent civilization wrote a Russian scientist, a leading researcher at Vinogradov Institute of Geochemistry, Ph. doctor of geological and mineralogical sciences Evgeny Ivanovich Vorobyov (1940 – 2006). He was a well-known specialist in the field of geological and mineralogical research, the author of many scientific papers and copyright certificates. Evgeniy Ivanovich has repeatedly personally participated in expeditions to the Irkutsk region, Buryatia, Yakutia, Krasnoyarsk Territory. Vorobyov believed that many underground cavities were of artificial origin and arose long before the Oligocene [25].

POSSIBLE NESTING AREAS

Underground caves of Cappadocia

Water sources, rich microflora and even flora and fauna reside present underground. Large, air-filled cavities exist below the surface of all continents. Their maximum depth does not exceed 5 kilometers, because at greater depths, the pressure crushes any voids. Very often such cavities are found under hills and mountain ranges. However, there are also underwater underground voids. Most of underground cavities have a natural origin, these are karst failures, lava tubes, etc. But some are clearly artificial and very ancient. In some places, traces of the presence of intelligent beings have been preserved.

Extensive underground structures have been discovered near Dirinkuyu and Kaymakli villages in Turkish Cappadocia. This is a whole complex, consisting of 36 multi-level underground settlements. These caves have long been uninhabited. For some time people lived there, apparently using ancient caves abandoned by someone in more distant antiquity. But these underground structures were not built by human people, a clear resemblance to termite mounds can be seen on the model of such a structure.

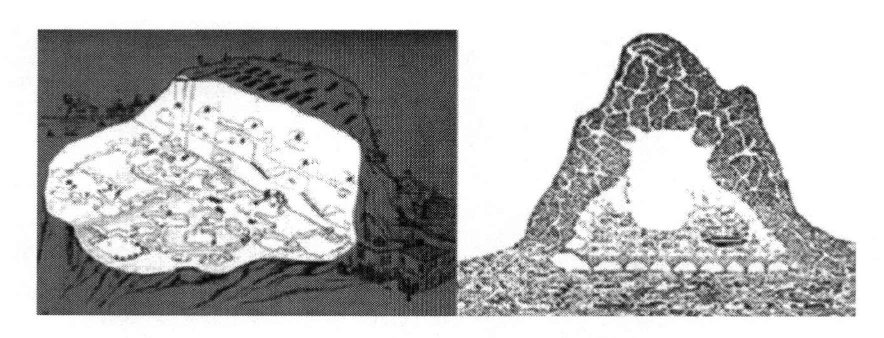

Spatial model of Kaymakli underground city (left) and a common termite mound (right)

Apparently, it also happened vice versa, underground inhabitants could use not only natural cavities, but also underground structures that were created by human hands, abandoned mines and catacombs. Sometimes strange creatures were found in the dungeons, and even observed in the networks of underground sewers. For example, in England, not so long ago, the camera of a cleaning robot filmed a strange jumping creature.

It must be assumed that many underground spaces are still inhabited, somewhere there, in depths of the earth, there are nests of intelligent insectoids living side by side with their symbionts. Underground roads and tunnels are laid between nests. It is not possible to list the known huge underground caves. Therefore, we will focus only on those could have signs of anomalous inhabitants. The close presence of nests and portals may be evidenced by frequent UFO sightings, suspicious disappearances of livestock and people, and by unknown radio emissions. By the way, it was precisely because of such electromagnetic anomalies that the expeditions of Barchenko and Bokiy to the Kola

Peninsula were organized back in the 20s of the last century (see above).

One of these suspicious places is located under mountain Monte Perdides in the Pyrenees to west of Andorra, on the border of Spain and France. There is a source of unknown radiation, and UFOs often appear over mountain. Ufologists suggest there is a large «base» of Aliens [26]. Very similar zone locate in Serbia above Mount Rtanj. Aircrafts, fliing above it become invisible to radar. Serbian physicist Ljubo Restovsky suggests that under Mount Rtanj there is an electromagnetic radiation source about 28 kHz. The presence of this source also explains magnetic and gravitational anomalies occurring within a radius of approximately 200 km from Rtani, due to which electrical appliances fail. In the mountainous regions of the Carpathians, belonging to Poland (Sandy Mountain), Slovakia (Mount Babya), strange traces were observed, people disappeared, and UFOs were also seen there [27, 28]. Polish explorer Jan Paenk reported huge underground cavities in these areas. Sometimes miners, while driling shreks, stumbled upon tunnels, walls of which were covered with solidified molten rock, similar to glass [29].

Dungeons stretched near Odessa and its suburbs for 2.5 thousand km. Odessa catacombs have many hide entrances. It is believed that most of the catacombs are former quarries. However, some tunnels are of natural origin. The same ancient catacombs were found on other side of the estuary, under the Akkerman fortress, near Belgorod-Dnestrovsky. It is possible that Odessa and Akkerman dungeons are connected to each other by tunnels. Strange «horned» creatures have been repeatedly seen in vicinity of the fortress. For some time now, underground explosions have been carried out there in order to block entrances [30].

Powerful underground structures, apparently, exist in the Crimea. In the mid-1950s, a glass-lined tunnel was discovered off Cape Aya, filled with concrete and refurbished, and now houses a military base known as «Object 100». Tunnels with walls covered with glass were found near Yalta, entrances to them were also concreted. Apparently, the dungeons are inhabited till now. People who were resting in village Sanatornoe near Cape Faros saw one of rocks move and two flying discs flew out of resulting opening (1971). The shepherd, who was standing on the eastern slope of Kara-Dag mount, near Rodnikovskoe village, saw a UFO that approached Mount Chavush-Kaya. A camouflaged gate opened in rock and the object entered smoothly, after which the gate closed. The portal was never found later (1990). In different regions of the Crimea, near Ayu-Dag, near Chatyrdag, at foot of Mount Boyka, on Karabi plateau and in other areas, people heard technological noises, as well as vibrations, which may indicate some kind of technological activity [31].

Extensive underground labyrinths of Kobyakovo settlement (see above) are located in Rostov region of Russia. They are very long and branched and look like artificial ones. The tunnels were discovered in Stavropol region near the Mount Strizhament, in the Izobilnensky district, in village Naidenovskaya. People and cattle also disappeared in these places [32]. In the Krasnodar Territory, not far from Gelendzhik, there is a suspicious bottomless well with a diameter of about 1.5 m. Speleologists lowered a video camera into the well to a depth of 200 m, but it showed only smooth walls. The radiation background at such a depth exceeded the natural one by 4-5 times. The researchers managed to record infrasounds, their intensity increased with depth. Experts who studied the mine suggested that it could not be a creation of nature or a product of human civilization. People actually disappeared near this place [25].

Significant underground cavities are also located under the Zhiguli mounts, in Samarskaya Luka area. One day, getting lost in these caves, a local amateur caver stumbled upon something like an incubator or columbarium. After passing through many branched half-littered underground passages, the unlucky adventurer got to a dimly lit corridor with smooth walls. There, in glass boxes that looked like blocks of ice, strange non-human monsters stood motionless [3].

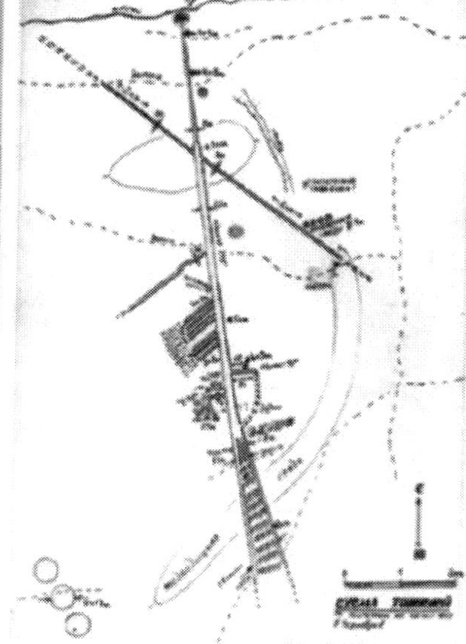

Medveditskaya ridge and the scheme of tunnels (by V. Chernobrov)

According to the data of research group from Togliatti led by Tatyana Makarova, there is an increased UFO activity and in Samarskaya Luka area, as well as various anomalies light. The researchers suggested that in this place there is a very large large intersection of underground roads, eight tunnels diverge from here [33]. A similar intersection node is located under the Medveditskaya ridge in Volgograd region. UFOs often appear here too. Since 1997, Kosmopoisk expeditions had studied in detail the underground voids beneath the area.

The Kashkulakskaya cave («black devil's cave») is located under the mountains of the Kuznetsk Alatau. Numerous experiments conducted by scientists from Novosibirsk made it possible to detect constant fluctuations in the strength of electromagnetic fields in this cave. There were different fixed frequencie signales, one of them had a stable amplitude. Sometimes it disappeared for several days, and then reappeared. It was found that the source of the appearance of impulses is located somewhere in the depths of the cave. Experts came to the conclusion that the signals from the Kashkulak cave are of artificial origin. However, the search for the source of radiation yielded nothing, since the emitter was in inaccessible depths [12]. Similar underground signals using the MTU-5 device were also detected by N.V. Sokulina and N.P. Nazarova near village Okunevo in the Omsk region (2004). They, too, were clearly artificial in origin. It should be noted that there are no military or industrial facilities nearby that could serve as their source [34].

Studies of tunnels, passing under the Putorana Plateau in the north of Krasnoyarsk Territory, were carried out by very serious scientific and leading organizations of the USSR. There, on September 29, 1975, an underground nuclear explosion was carried out at a depth of 800 m, allegedly with the aim of creating underground gas storage tanks. Two years later, a new nuclear explosion was carried out a kilometer from the first one. However, these «capacities» were not used in this way [35].

In the 1950s, the USSR began to build an underground tunnel connecting Sakhalin with the mainland. The builders were laying a modern tunnel along the ancient one, laid by unknown builders. When mysterious objects similar to mechanisms and fossils were found in the bowels of an ancient tunnel, everything was immediately classified [25]. Perhaps for this reason, the project was closed, although billions of rubles had already been invested in it, and it was very necessary for the development of the Far East.

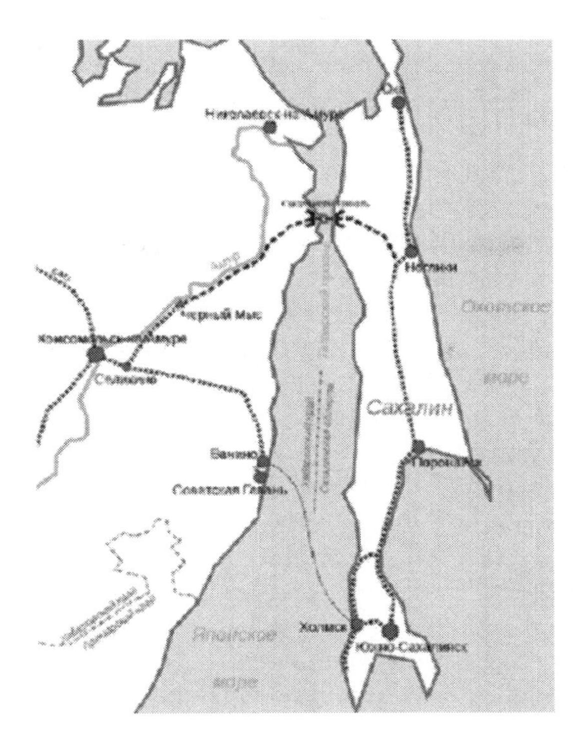

Tunnel connecting Sakhalin with mainland

There are underground facilities in the foothills and mountainous regions in the east of Kazakhstan. Abandoned «nests», underground cavities look like giant, smelly and dirty «termite mounds». Sometimes they were used by people to create underground settlements. There are extensive caves in the Altai Mountains and in northern Iraq, inside the Kun-Aun mountain range in China. On the border of China and India, in the Ladakh region near the Kongka La mountain pass, too, according to the Indian journalist Sudhir Chad, who published in the India Daily newspaper in 2013, there is an underground nest of Aliens. Pilots of the Chinese and Indian Air Force avoid flying there and diligently fly around the area. According to them, at first navigation instruments turn off, and later, if plane does not turn back, engines fail. Here, too, witnesses sometimes see UFOs.

A large nest is located under Mount Hayes in the Brooks Range, northeast of Anchorage, Alaska. Over this area there are constant failures of electrical appliances of aircraft and spacecraft. This area is generally known as the «Alaska Bermuda Triangle»: not only people disappeared there, but also planes. Military planes from the Eielson Air Force Base, located 42 km southeast of Fairbank, allegedly tried to bomb this alien nest, but the «answer» came from there [36].

Apparently, an Alien nest is located in the foothills of the Blue Ridge under Mount Brown, near Morganton in North Carolina. Strange phenomena and fires which used repeatedly observe there, were photographed by two researchers from Appalachian State University, Daniel Caton and his colleague Lee Hawkins [37].

The most intriguing mystery seems to be hidden in southern Colorado under Archuleta Mesa. There, not far from Dulce, is the Jicarilla Apache Indian Reservation. Beginning in 1947, there was some building of a road and a logging enterprise. However, the logs were never removed from there, and the road was subsequently abandoned. Dr. Paul Bennewitz, known American physicist, located an underground Alien's nest under Archuleta Mesa. He registered some radio signals coming from under the mountain. According Bennewitz, underground passages from under Archuleta Mesa stretch for many kilometers and connect with natural cave systems. There are many cases of animal mutilation, as well as cases of kidnapping. Perhaps the event when during underground construction work many soldiers died and Philip Schneider was badly wounded happened right here, near this nest (1979, see above) [15, 16].

It is believed that one of the nodes where underground paths converge is Mount Shasta, a dormant volcano belonging to the Cascade Mountains. American researcher William Hamilton devoted a lot of time to its study [14]. The famous British traveler Percy Fawcett also believed that the tunnels stretched from Mount Shasta to the volcanoes of Popocatepetl and Inlaquatl in Mexico. His opinion was based on the stories of local Indians. Many mutilated animals, ever bisons, have been found not far from the mountain [9].

In 1934, a geophysicist, engineer Shufeld used his patented device to explore underground cavities located under Los Angeles and discovered a whole network of tunnels [38]. Some nests or bases are located on the island of Puerto Rico. In 1980, police officer Ivan Rivera Morales allegedly went inside an underground structure located under the El Cahul mount in the Laguna Cartagena area. According to him, there was a large industrial complex here [39].

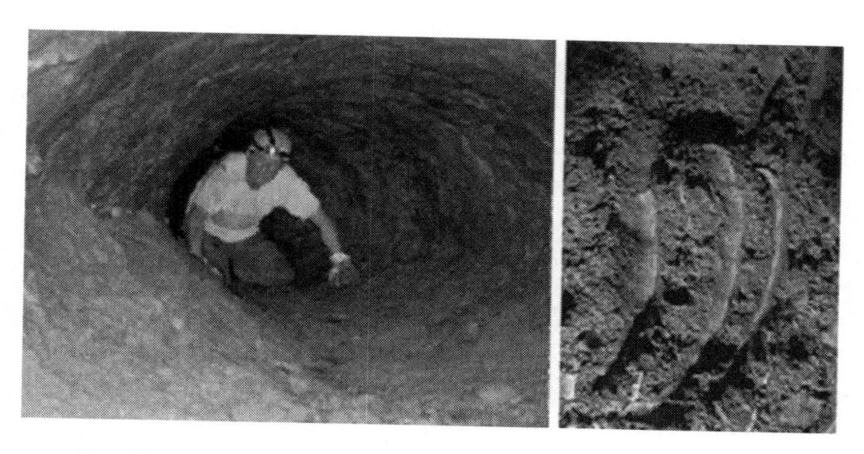

Brazil underground tunnels and a claw marks on a walls

The Mapimi Bolson Biosphere Desert Reserve located In the Mexican state of Chihuahua. There is a so-called «Zone of Silence». This is an anomalous zone covering approximately 50 square kilometers. Any radio signal is jammed here, electrical appliances, compasses and watches do not work. The first time this have been noticed in 1934, when pilot Francisco Arabia, lhaving lost orientation, had been forced to make an emergency landing. In 1964, chemical engineer Harry de la Penia tried to figure out causes and nature of anomalies. Howewver it is unlikely that this was due to ore deposits, as the location of the anomaly have been shifting. Strange lights and fly objects have been repeatedly observed over this area, one have been photographed over Magnetic Mountain (1976) [40].

Mysterious underground underground cavities localise in Mexico, the largest of which is the Cave of Swallows. Its walls are absolutely even and smooth, and the depth is more than 1 rilometer. There are labyrinths and rooms and passages that diverge in different directions.

Geology professor Heinrich Frank discovered many such caves and tunnels in the South and South-East of Brazil in 2016. These underground passages was branched, sometimes chambers are separated from them. On the surfaces of walls, consisting of weathered rocks of granite, basalt and sandstone, they have characteristic parallel furrows resembling claw marks [41]. Interestingly, approximately the same traces are present on the walls of the underground cities of Turkey.

One of these tunnels, called Posid, is located in the central Brazilian province of Mato Grosso. Another huge cavity in the neighboring state of Rondonia was explored in 2010 by geologist Amilcar Adami of the Brazilian Geological Survey. He concluded that these structures were not a result of any natural process. Entrances to these tunnels are located in the states of San-Paulo, Parana and Santa-Catarina. Meanwhile, high anomalous activity is observed in Santa-Catarina, local residents have repeatedly heard underground technological noises there.

The most famous networks of underground cavities in South America are found in Ecuador and Peru. In 1965 Argentine ethnologist Juan Moritz discovered a tunnel net in the Ecuadorian provinc of Morona Santiago (he reported this to President of Ecuador in 1969). The entrance to the dungeons was carved into the rock and led to a depth about 240 meters. Tunnels were very long and looked like labyrinths.The surface of walls and ceiling was smooth, as if polished. Ventilation holes in the walls, located strictly periodically. The underground complex included horizontal platforms, successively located one below the other. The tunnels diverged from them in different directions.

Even more impressive dungeons, the so-called «Chincans», were discovered in the Peruvian Andes. In the 1550s, Spanish conquistador Francisco Pissarro informed Spanish King about the discovery of a dungeon under Mount Guascarane. The entrance was at a high altitude, and it was impossible to get in underground tunnels, because the entrance was blocked by stone slabs. There, underground passages made their way in hard rocks, and today they are practically not explored. By order of the authorities, all entrances to them are tightly closed with bars. In Chinkanas, dozens of adventurers and several research expeditions have already disappeared without a trace, incl. from Lima (1923) and the French expedition (1952). The researcher of the Inca civilization, Dr. Raul Rios Centeno, with a team of speleologists, tried to penetrate these dungeons, despite the prohibitions of the authorities. They still managed to get into the underground tunnel through one of the entrances, located a few kilometers from Cuzco. However, the tunnel gradually narrowed, and when movement became impossible, the group returned back (1971). In 1991 Peruvian speleologists discovered a network of caves in the area of the Rio Siju River. At a depth of 70 m, a stone slab blocked the road, but they managed to move it, it rotated around its axis. Behind it was found a tunnel going down. Two parallel grooves were carved along the floor of the tunnel, as for a wagon. An expedition in 1995 discovered that this tunnel goes under water [20, 42].

Thus, there is reason to believe that there is enough space in the underground cavities for an unknown species to live. Its real existence there is evidenced not only by anomalous phenomena near the portals, but also by clearly artificial processing of walls, technological noise, and strange electromagnetic radiation with fixed frequencies. It is not always possible to explain them by the proximity of iron ore deposits,, any mines or military installations.

NOTES of PART 2.

1. Extraordinary Commission, People's Commissariat of Internal Affairs, United State Political Administration – in different years they performed the functions of USSR state security.

2. Shishkin OA «Battle for the Himalayas. NKVD: magic and espionage», 1999 (in Russian)

3. Pavlovich IL, Ratnik OV «Secrets and legends of the Volga dungeons», 2003 (in Russian)

4. Pervushin A L «Occult wars of the NKVD and SS», 2004 (in Russian)

5. Ivashov L G «The overturned world. The secrets of the past are the mysteries of the future. What the archives of the Special Department of the NKVD, Ahnenerbe and the High Command of the Wehrmacht hide», 2016(in Russian)

6. Maccabee B «UFOs and the FBI: US Government X-Files», 2001 (in Russian)

7. Corso Ph «The day after Roswell», 1998

8. Stringfield LH «UFO Crash/Retrievals… Status Report III» 1982-1991 cited in Агон Е, Анфалов А «Небесные ангелы или псы ада», 1922

9. Howe L «An Alien Harvest. Further Evidence Linking Animal Mutilations and Human Abductions to Alien Life Forms», 1989

10. Cooper W. «The secret government. The Origin, Identity, and Purpose of MJ-12»// Quest Publications International Ltd, 1989/1990

11. Sauder R «Underground bases & tunnels: What is the Government Trying to Hide?» Paperback – February 20, 2014

12. Voitsekhovsky AI «Secrets of Atlantis. Great secrets», 2000 (in Russian)

13. ren.tv/project/samye-shokiruiushchie-gipotezy/841791-podzemnyi-mir-samye-shokiruiushchie-gipotezy-s-igorem-prokopenko-03-06-2021

14. Hamilton W «Alien Magic» 249 North Brand Boulevard, Suite 651 Glendale, CA 91203

15. Branton (Author), Beckley (Editor) «The Dulce Wars: Underground Alien Bases and the Battle for Planet Earth», 2011, 135 pages

16. Darcy Weir, «The Undergrround: Director's Cut», 2020 – Filipp Schneider managed to give 30 lectures, many assassination attempts were made on him, as a result, he was killed on January 17, 1996. All the materials he was preparing for publication disappeared without a trace.

17. «Did the USSR develop weapons against aliens?» – http://paranormal-news.ru/news/v_sssr_razrabatyvalos_oruzhie_protiv_inoplanetjan/2013-09-10-7654

18. . Paulides D «Missing 411. Western United States & Canada : unexplained disappearances of North Americans that have never been solved», 2011

19. Nifontov BI, Protopopov DD, Sitnikov IE, Kulikov AV «Underground nuclear explosions», 1965 (in Russian)

20. Burovsky A «Reason and Civilization, or Flicker in the Dark», 2017(in Russian)

21. «What was it: an ultra-deep well after 12 kilometers of drilling reached hell» - news.rambler.ru/other/44914382/?utm_content=news_media&utm_medium=read_more&utm_source=copylinkhttps://news.rambler.ru/other/44914382-chto-eto-bylo-sverhglubokaya-skvazhina-posle-12-kilometrov-bureniya-dostigla-ada/

22. Loki Wotan «Tales of Grandfather Shaman. Evil Spider» -www.youtube.com/watch?v=x4bqsDaYgmA&list=PLGWsC6thRojkuZ6jVBhXmSM8A5wS3GXQn&index=6

23. Volkov E. «Did not return from the campaign» // TOURIST. – 1973. – No. 6. – S. 12-13 (in Russian)

24. Prokopenko I. «Guests from space. Facts. Evidence. Investigations», 2019

25. Vorobyov E, «The Underground World» // ORACLE, No. 3 (132). 2005 (in Russian)

26. «Project 8200 Declassified | The Army Tried to Analyze If Ancient Subterranean «Relays» Existed»- www.youtube.com/watch?v=H4AvcCAcs_k

27. Fort CH «Lo!», HYPERTEXT EDITION BY MR. X – www.resologist.net/

28. Dash M «The Devil»s Hoofmarks source material on the great Devon mystery of 1855», FORTEAN STUDIES ACADEMIA.EDU, 1994, 71-150

29. Prokopenko I «Secrets of natural anomalies», 2022

30. Lyakina E , SECRETS OF THE USSR, June 2021, No. 10 (in Russian)

31. Anfalov AA «Aliens in the Crimea», 2018 (in Russian)

32. Museum of Local Lore, Road to Hell, BLOCKNOTE, STAVROPOL, 28.05.2018 16:17 – bloknot-stavropol.ru/news/-doroga-v-ad-ili-kakie-zhutkie-podzemnye-tonneli-e-977798

33. Makarova T «Chronicles of anomalous phenomena. Notes of a thinking observer». Vol 1, 2, 2016, 2020 (in Russian)

34. «In touch. Extraterrestrial civilizations. The world's first hardware contact» – youtube.com/watch?v=DWB0YcMPuc8

35. Struchkov S «Peaceful nuclear explosions of Taimyr»// GO ARCTIC, August 13, 2018 – goarctic.ru/work/mirnye-yadernye-vzryvy-taymyra/

36. «Missing in Alaska» – www.youtube.com/watch?v=-eDtwmy-MXM ; www.youtube.com/watch?v=Xt9JJjfZk88

37. «ASU scientists think they»ve captured images of WNC»s unexplained Brown Mountain Lights» – wlos.com/news/local/asu-scientists-capture-rare-images-of-wncs-brown-mountain-lights

38. Los Angeles Underground «Lizard People» Tunnel Map according to engineer G. Warren Shufelt – LA TIMES, Jan 29, 1934

39. «Alien bases in Puerto Rico» – earth-chronicles.ru/news/2016-12-16-99404

40. goaravetisyan.ru/zona-molchaniya-v-meksike-anomalnye-zony-zemli-zona-molchaniya/

41. Karachev G «Hundreds of tunnels created 10,000 years ago were discovered in Brazil»// The Russin Time, 6.06.2021 – therussiantimes.com/istoriya/413075.html

42. Azhazha, V.G., Litvinov «Underwater UFOs», 2015 (in Russian)

PART 3
CONTACTS

FORAGER CHUPACABRUS

Among insectoids, most often people encounter hunters-foragers who rise to the surface in search of prey. They supply Alien nest entire habitants with nutritious foods. A typical representative of this caste was described in part 1 of this book. The authors of REPORT-96 called him «Small Worker», although there is no certainty that he should not be included in a caste of fighters. The blood of warm-blooded animals is a rich source of nutrients and iron, which aliens need in large quantities, since it is a necessary component of their magnetosensor organs and chitinous covers. However, not all foragers are vampires, but absolutely all of them are predators capable of nibbling and tearing their prey to shreds.

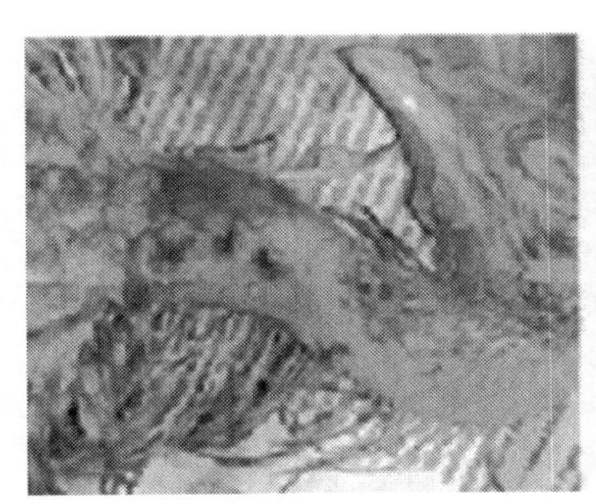

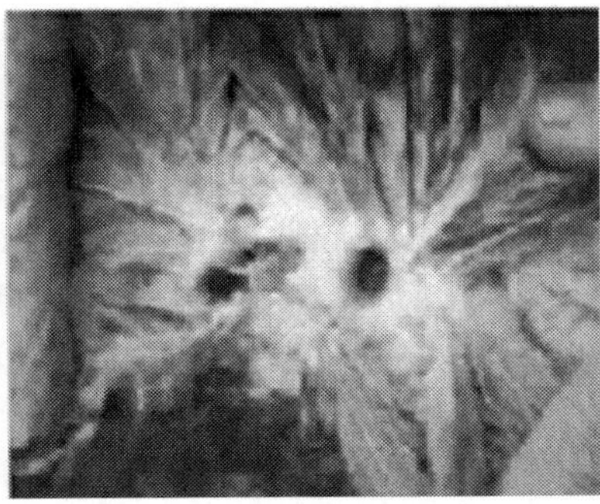

Typical Chupacabrus bites

In Latin America, Europe and Asia, these creatures are called «chupacabra». Other well-known monsters can be attributed to Chupacabrus. Jersey Devil has terrified farmers in North America since the late 18th century. In the same place, at different times, Gotman and Motman appeared. In the Far East, animals are attacked by Flying Man. It is possible that the legendary British Jumping Jack, French Beast of Gévaudin and Czech Perak were also representatives of this group of animals. One gets the feeling that a differences between them are determined not so much by their actual characteristics, but by a difference in perception of witnesses from different ethnic groups (Latinos, Anglophones, etc.), according to their traditional culture and customs. The analysis of testimonies was carried out in different regions of the world: in Western Europe, North America (USA, Canada), Latin America, Eastern Europe, Northeast Asia and other regions [38]. Therefore, we will call all these animals «Chupacabrus».

The absolute number of witnesses considered Chupacabrus to be «skinny» upright animals, 1.5 – 1.7 m tall, with long crooked legs, short forelimbs, a large head and an elongated muzzle. In Chupacabrus, in comparison with representatives of other castes, a «muzzle» seems to be longer due to a specific structure of their oral apparatus. Perhaps that is why they are sometimes compared to «upright wolves». Animals have cancer claws shaped feet and upper paws with paired claws. These predators leave traces that are located «in one line.» They look like the footprint of a bird or a cloven hoof. Running away, Chupacabrus was jumping on two «curved» legs, outwardly resembling a kangaroo or a grasshopper. At the same time, he looks higher or lower, depending on how deep he «squats». And absolutely everywhere Chupacabrus in the dark shine with their huge red-yellow eyes without pupils [38]. In almost all regions of the world, Chupacabrus could fly, had wings or some kind of formation on their backs: a hump, a mane, a cape, spikes, which, most likely, were also folded wings. The striking similarity of Chupacabrus from different regions of the world is manifested not only in their appearance, but also in their habits. So despite some, not too significant differences, the relationship of the Chupacabrus of all continents is beyond doubt.

Chupacabrus are not only unusual, but also extremely ancient animals. Some researchers are trying to explain the appearance of Chupacabrus by genetic experiments conducted on US military bases. However, these creatures appeared long before any bases and people. These bipedal creatures have been leaving footprints on the ground since the time of the dinosaurs. «Cup» imprints of bipedal creatures have been preserved on exposed rocks on almost all continents of the world [26]. They are tens of millions of years old. Although some researchers are trying to attribute them to ancient people, it should be emphasized that at the end of the Jura there could not be a person on Earth, and primitive small mammals only recently split into marsupials and placentals.

Bloody attacks of Chupacabrus, such as harpie's, have been preserved in prehistoric memory among different peoples of the world. The appearance of «vampires» in medieval villages was accompanied by loss of livestock and panic. Chupacabrus often get close to human habitation. In REPORT-96 there is a mention that at night small workers go out into a «open space» and walk «among the buildings». They operate in dark only, afraid of lights. In case of danger, «these animals tend to be in the least lit place». They are ready to run at any moment. They reaction to a strong stimulus consists in a jump «up and to the side.» Just hiding from a light and crouching to ground, Chupacabrus used to appear, for example, in Eastern Europe, in Puerto Rico (2019) or in Chile (2000). Like Small Worker from REPORT-96, Chilean Chupacabrus saw teenagers, jumped high, and then He has hidden under a standing pickup truck (2001) [27-31].

Not only anomalist researchers, but also zoologists and veterinarians noted that Chupacabrus are completely unusual predators. This was stated in Mexico by one of veterinarians Dr. Soledad de la Penia (1996) and a biochemist Dr. Marco Reinos (2000) [32]. In the spring of 2009, village of Zastavnoye head, of Lviv region, Nadezhda Rudaya, after examining remains of dead animals and traces left by a predator, said: «I have livestock veterinarian education and I can say that we are faced with something very strange...» Zoologist Oleg Poyasnik from Chernihiv devoted several years to studying a strange phenomenon: «When people started talking about attacks on pets in our area, I was skeptical about it. I decided to get acquainted with facts in order to prove that there was nothing supernatural. I could assume only that this is a predator, which is atypical for our places. But what I saw did not fit into the zoological framework ... Rudiments of intelligence are clearly inherent in him, but in a form that a person cannot yet understand...» Referring to unusual concomitant circumstances, he said: «I am a zoologist. If Chupacabra does exist really, I suspect it is not a zoological phenomenon» [33-35].

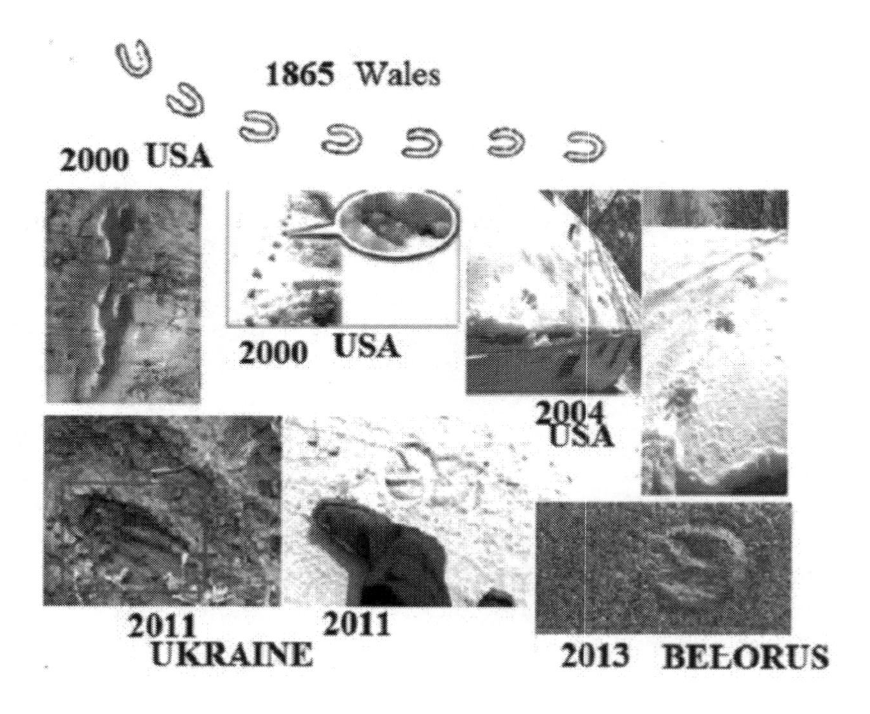

Devil's traces from Wales (1864),
Devil's traces from New Jersey (2000, 2004) and
Chupacabra's traces from Eastern Europe (2011, 2013)

The characteristic features of vampire attacks were long overlooked and attributed to completely different animals, such as wolves, which often «slaughter cattle» in excess, leaving intact carcasses with wounds on neck. Bloodletting of chickens, rabbits and other small domestic animals, could be attributed to predators of a weasel family. Anomal phenomenon began to be noticed theonly after development of communications, when a wave of rumors and panics swept across continents. However usually local authorities around the world are doing everything to keep this phenomenon «unmanifested».

The foot of the Chupacabrus is like a cancer claws, so its print is often mistaken for a cloven hoof. Sometimes additional processes could be printed between the two claws. Such prints were repeatedly seen: in Stavropol Territory (January 2011), in Kharkov (September 2011), in October – in Buzovaya village of Kyiv district (October 2011), in village Gora of Kyiv district (June 2013), in September, in village Gremyache of Voronezh district, and in Bayangol village (or Sael) of Barguzinsky district (September 2013). In 2015 similar tracs were fund in Somovo village of Voronezh region. Exactly same imprints were left by Devils from Jersey in the USA and the Great Britain.

These monsters hunt attaking small domestic and wild animals. They attack unexpectedly from ambush or jumping from trees or walls, they attack elso by running of flying. They jump over fences, break open cages, crushing wood and breaking metal using their upper limbs equipped with pincers, pointed like claws, up to 8 cm long. Bloodless bodies with wounds on a neck, less often on other parts of body are left after an attack of vampires. As a rule, a corpse does not decompose and does not have a characteristic rigor mortis.

The study of bloodless remains of victims showed that their wounds are completely different from those left by known predators. Most often, two holes are found, their depth can reach 9-12 cm. Thay are track of the long mandibuls. Less often, there are three holes located in triangle: two large holes and a smaller one lower and in the middle. Such a bite pattern is obtained when upper mandibles and lower maxillas dig into a victim's flesh (usually lower holes are double, but hardly noticeable). A powerful circular muscle is located around a mouth opening, which creates a high negative pressure for the absorption of fluid. Witness Lena Barkhotina from Kremenchug saw the Chupacabrus raise his «upper lip» before alleged attack [36].

The attack scenario looks something like this. First, the victim is struck with a blow that does not kill, but only immobilizes her. By biting through a skin and muscles, Chupacabrus inject into the victim's bloodstream a gland secret containing anticoagulants and hydrolytic enzymes. Enzimes that were found in Small Worker digestive tract have a pH optimum in a slightly alkaline or neutral aria. Thus, they can be quite active in the bloodstream, because animal plasma pH is 8.4. While the heart is working, anticoagulants and fibrinolytics are easily distributed throughout the body of an animal. It takes some time to complete it, so the Chupacabra puts the first victim aside and takes a next one. The vampire carefully lays out the carcasses to suck out liquids in the same order after a short time. In some cases, the monster is unable to complete the blood sucking by some obstacles. Then a victim remains alive, but with profuse bleeding.

There is reason to believe that the purpose of the attacks on the Jersey Devil and Jumping Jack was also to suck blood. In any case, after the attacks of the first, animals remained dead but not eaten, and after the attacks of the second, some victims remained alive, but with profuse bleeding. (Although more often Jersey Devil, Mothman and Gotman tore victims to pieces. However, all chupacabrus are ables do the same).

According to REPORT-1996, a stomach volume of Small Worker is quite impressive, about 20 liters. During the hunt, the gluttonous creature does not calm down until it is satisfied, destroying dozens of rabbits and other living creatures. The stomach serves for partially digestion and for depositing liquid food. Having filled it, it is difficult to fly, so the burdened monster goes to a nest by jumps. There, the forager regurgitates the prey in the form of a semi-digested mass. All other inhabitants of the family feed on it. Regurgitation is stimulated by pheromones that release of from Upper cast individuals.

Hunting Chupacabra foragers is an ancient natural way of obtaining food for nest inhabitants. (The second method is much more effective. Hunting is carried out from UFO board, while large animals become victims: horses, cows, elks, deer, etc. See below).

The insectoid nature of Chupaquibras has ample evidence. Some residents of Puerto Rico (1996), Mexico (1996) and the South Urals (2017) have noticed a hard exoskeleton of Chupacabrus, «like a cancer shell». A witness interviewed by Nick Refern said when he has heard a noise close of his house, he quickly ran out and tried to hit a monster with a machete. However metal tool ran into contact with hard armor. A farmer who regularly butchers pigs with the same machete was surprised [37]. But sometimes it seemed that a monster body was covered with something like an cloak or coat (may be it was wings).

Almost all varieties from different regions of the world had wings or were able to fly. Even in Eastern Europe, where there is little direct observation of flight, people have seen flying monsters. It seemed to Lena Barkhotina from Kremenchug that the strange creature had something on its back that resembled the *double* wings of a dragonfly (1990) [36]. Sometimes these creatures soared vertically upward, leaving a trail of steam behind them. For example, in Poland, peasants watched the takeoff and flight of Chupacabrus (2004). Although if we draw an analogy with termites, then we can assume that workers and soldiers may not have wings at all. Sometimes these creatures soared vertically upward, leaving a trail of steam behind them. Sometimes, on the contrary, they took off heavily, with a running start, like a low-flying chicken [27]. In Barcelona, an eyewitness saw a similar creature which he called a «gargoyle», take off from the balcony balustrade and fly. It was a full-fledged flight, although the creature flew hardly [48]. Some witnesses have seen Chupacabrus and Jersey devils inflate their crops (air sacs) before taking off [38]. Probably, before flying, the Chupacabras needs to saturate the blood with oxygen. Perhaps it is easier for a creature to fly on an empty stomach. Zoologist Oleg Poyasnik noted that at the hunting site, the creature's traces appeared as if from nowhere, but usually after saturation, the vampire ran away on foot [33].

It is very important that, as many eyewitnesses noted, the wings of the Chupacabrus are behind the back, and the upper limbs are in front, completely separate from the wings. That's why these animals cannot be classified as winged vertebrates: birds, plesiosaurs, or flying mammals. In all these animals, wings were formed on basis of forelimbs. (The exception is some reptiles, for example, lizards of *Agamidae* family, or a flying dragon of the genus *Draco*. Their wings are formed on basis of ribs processes. But these animals are capable to do a short gliding flightof only). Chupacabrus, judging by eyewitness accounts, fly well. Thanks to their wings, Chupacabrus could made real flights, this ability is especially well developed in Flying Man of the Far East, Mothman and Jersey Devil.

It is interesting to trace the history of the appearance of Chupacabrus around the world.

To date, a whole archive of reports of encounters with the Jersey Devil, from the 18th century to the present day, has been collected by researchers from the US state of New Jersey [62]. Since the end of the 18th century, He has been seen many times on the east coast of North America. The year 1909 was especially rich in events. Outwardly and in habits, Jersey Devil is very similar to Chupacabra, although it may be a separate variety of them. (As we noted above, cases where Jersey Devil has sucked blood are not known, but may be investigations have not been done).

Since the beginning of the 19th century Jumping Jack, who, in our opinion, was also a typical Chupacabrus, has been hunting in Britain. According to eyewitness descriptions, he was tall, upright and «bony». He had bulging red luminous eyes, a black cloak or something similar to bat wings [56, 57]. Animals died in the countryside, and travelers died on the roads [38]. Chupacabrus traces were found in the snow in February 1855 at the mouth of the Exe River in South Devon (1855) and on Sand Mountain between Poland and Galicia (1865) [39]. Rural trackers reported many appearance of «Devil's footprints» in the form of a cloven hoof (cancer claws) [26,56, 59, 60]. Devil ran about 160 miles, while he overcame obstacles by jumping or by flying. Some peaple saw his flight and compared it with a bird [168]. In one case, he attacked a woman who was beaten off by gypsies who happened to be nearby. She watched him run into the darkness, jumping over bushes and hummocks. Another witness described his movements as those of a huge grasshopper [166, 167]. Later bloodless sheep were found in the British Isles, in Badminton (1905) and in the vicinity of the Windsor castle

(1906) [32, 40]. Attacks were typical of Chupacabrus: slaughtered rabbits and sheep had characteristic wounds and broken vertebrae (1910) [26].

In the second half of the 19th century, some strange creatures killed animals and people in different provinces of the Russia Imperium. In 1874, even military units were sent to search for them [40]. By the end of the 19th century in St-Petersburg, night watchman Georgy Semenov had been faced with Chupacabrus [41]. This phenomenon also took place in the Far East. In 1909, a peasant Ivan Zharkov from village Olemtsy in Irkutsk province (Yakutsk region now) has been encountered a creature very similar to Chupacabrus in his own yard [41]. In early February 1913, Chupacabrus had attacked small animals in village Lyakhovo near Kyiv (Osichki now) [42].

According to Lauren Coleman and Scott Corrales, Chupacabrus raids occurred long before the 1960s, they were called «Moka vampires» or «Comecogollos» then. In 1967, 1975 and 1985, there were cases of killing sheep and goats in Puerto Rico.

Chupacabrus «Owl Man» has made several appearances near the church cemetery in village Mawnan in Cornwall (1976, 1978, 1989) [61]. This was a huge creature with fiery eyes walked slowly, waddling. Then he flew up vertically, and everyone noticed his legs, similar to cancer claws (pincers). (However, the drawing by the witness, 15-year-old Perry, is inaccurate: creature's forelimbs and wings must be separated). In a couple of years bloodless sheep were again found in the British Isles in 1991, 1993 and 1995. Predators themselves were also seen there [25, 40].

This is how a teenage girl saw an Owl-Man
in Mawnan (1976)

Since 1979, chupacabras again have attacked animals on the other side of Atlantic. Strange deaths of domestic and wild animals occurred in Tenerife (1979) [38]. Many goats and sheeps died on the Iberian Peninsula: near Mount Espadas, in Ilso, Angostura, Antugnano, Urbas and Galicia. An unknown strange predator was also seen in Cinco Villa in Aragon (1979, 1996-1998) [32].

In the 90s, Chupacabrus continued to appear in East of Eurasia. In 1991, small domestic animals died due Chupacabrus attacks to Omsukchansky district farms of Magadan. The bodies of victims were not eaten, but veterinarian recorded two small wounds 5–7 mm diameter and up to 12 cm deep [32]. Since October 1990, hundreds of bled sheep and goats have been found on pastures in Calaveras County, California. Chupacabrus continued to appear in 1994, on Puerto Rico, not far from Orocovis, in Guanica and Lakhas. In March 1995, poultry farms of Florida were damaged, and in early May, another murder occurred in the same state. In Highley Gardens, a dozen and a half goats and sheep were killed [43]. In 1995, the animals died elso in Houston County, Texas. Soon Mexico was got under attack. To evoid panics, Mexican government imposed censorship on publications. The bans did not change anything, and on May 3, 1996, the Chupacabrus have attacked a sheep farm in village Puebla.

Next years, vampires flourished in Central America: Guatemala, Panama, Costa Rica, El Salvador, and Venezuela. Later, the phenomenon spread to South America (1996 – 1999): northern Chile, near Calama and Concepcion, in Paraguay (1998) and Brazil (1999). In 1998, Vittorio Pacacini reported that Chupacabras has been notised near of Sete Lagoas in Brazil at a depth of 150 meters in a limestone cave [63].

In 2000 after repetedly killing domestic animals in near Calama, Chilean police and military incl. NATO troops, joined to the hunt for killers. In April and May, vampires were bombed in the Atacama Desert. Two animals were allegedly killed and one taken alive (they probably have been shipped to the US).

Perhaps because of these hostilities, the phenomenon gradually moved to south. Since end of April 2000, animals have been dying already in Central Chile, not far from Santiago. On April – May animals were killed and bled in Arauco and Laguna Redonda [30, 38, 45].

In 1998, a Chupacabrus having killed a kangaroo was seen in Australia's Yanchep National Park [31].

In September of the same year, anomalous attacks were reported from Polish Radom. There, at the edge of the Kozienicka Forest in the forest, two people met a strange monster and three dead goats were found with signs of typical Chupacabrus exsanguination [44].

Strange animals appeared in the Crimea elso. Morning, January 7, 2002 in village Beloglinca, local residents saw classic U-shaped footprints of a bipedal creature coming from the cemetery. Some times the tracks were interrupted and then could be found on the roofs, as this creature flew through the air. On

April 22, 2006, these strange footprints were seen again on melted snow in Mount Kemal-Egerek. On August 2003, for several months habitants of Novoandreevka and Kharitonovka villages reported about a suspicious «horse woman». Instead of legs, she had hooves, her body was wrapped «in a long black fur coat». Needless to say, this portrait fits to description of Chupacabrus, specifically British Jumping Jack (1937) or American Jersey Devil (1778 – 2016). By the way, the latter was often mistaken for a two-legged horse or deer [38, 64].

After 2003, reports of Chupacabrus attacks came from Belarus, the Ukraine and more than two dozen administrative regions of Central Russia, as well as from Southern Urals, Altai Territory, Novosibirsk Region, from the former Central Asian republics of the USSR, as well as from the Far East [38]. (Due to the lower population density, there were fewer reports from the eastern regions).

After 2001, again hundreds of cases were reported involving Chupacabrus: from Argentina (2002), Colombia (2003) and from the United Kingdom. In April 2007, one Staffordshire newspapers reported about strange «upright walking wolves» with have been were repeatedly seen between Stafford and Cannock. Other hand on April 5, 2010, English newspapers reported a new mass loss among sheep [46, 47]. What looks like a Chupacabrus was filmed by a cleaning robot in England. The video clearly shows how something resembling a kangaroo is rapidly moving past the camera. Unfortunately, the details could not be considered.

On May 22, 2004, Chupacarus was spotted in Indian state Uttar Pradesh. Chupacabrus victims have been found in the Philippines (2008), Namibia (2009, 2012 and 2018), Vietnam (2010), China (2016) and Indonesia (2017). Many incidents occurred in Kentucky (2011), Ecuador (2013), Puerto Rico (2018) and Mexico (2019). It was not always reported whether corpses were bled, but if there were deep wounds, there were no traces of blood around.

In 2016 – 2018 Chupacabrus attacked cattle and people in India [52 – 55].

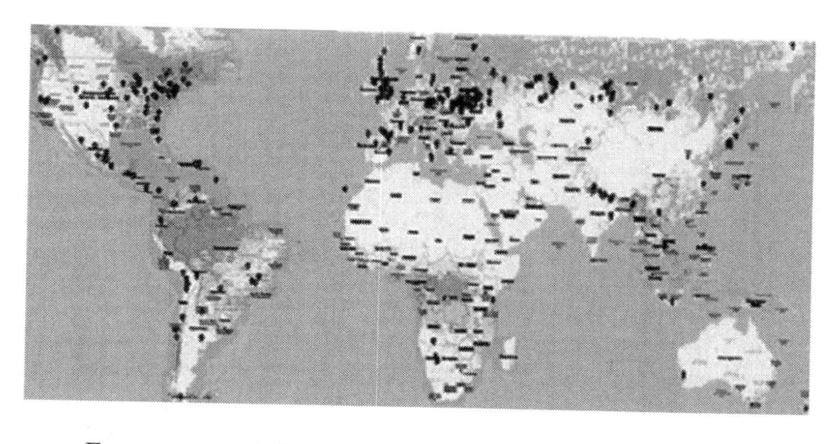

Emergences of chupacabrus over the past 200 years [38]

Chupacabrus attacks on animals continue to this day. On the maps of the two hemispheres, you can see what a vast territory these animals occupy. At the same time, there is a high probability that places not marked with dots are not zones free from them at all. It is quite obvious that no animal can have such a wide range of distribution, except, of course, humans and theirs domestic animals. Despite the vast distances and different climatic conditions Chupacabrus can appear anywhere in the world. Probably, such a distribution of places where monsters appear can be associated not so much with their habitat, but with fast movements. Although the Chupacabrus runs quite fast on its own legs, it is clear that even he cannot cover a distance of thousands. So the ability to fly and run fast cannot explain such long-distance movements of the Chupacabra from one area to another. On the territory of the former USSR and in Eastern Europe, Chupacabrus appeared in different places. The phenomenon moves chaotically, lingering for no more than a few days and constantly changing hunting grounds. Often hunting is carried out in several places almost simultaneously. For example, on March 22, 2012, vampires hunted on arias of Ternopil, Vinnitsa and Cherkasy regions of the Ukraine. Sometimes vampires return to an already verified place after a year or two.

Comparing the dates of incidents with each other, it is difficult to agree with a proposed Chupacabra dissemination scheme from Central America to the south and north, and from Western Europe to Eastern and further to Siberia and Central Asia [38]. These creatures appear almost simultaneously in very distant places from each other. The pattern in their course is barely visible, and the distribution of cases in space and time rather indicates the spread of rumors and panic among a population, which are reflected in the media.

It is clear that Chupacabrus move in some unusual way. Sometimes UFOs are observed in their hunting grounds, which land or pick up Chupacabrus from their «walking» places. For example, one lady driver watched Chupacabras being loaded aboard a flying object on a New York state highway near Niagara Falls (1958), a group of five people saw the same boarding in Texas (1953) [49, 50]. In Mexican villages near Abasolo, where sky lights were seen many times, strange upright creatures with red eyes appeared after object flying landing (1997) [32]. Appearances of Chupacabrus and UFOs have been coincided in time on Barsa-Kelmes island in the Aral Sea in 1985 [51] and in Sumy region of the Ukraine in 2010 [52].

Unfortunately, Chupacabrus attack not only animals. As a rule, these strange creatures run away at the sight of people. But some of these meetings ended tragically. Most of attacks took place at night or at dusk. Sometimes the monster jumped from a tree or from a wall, blocking to a passerby a path. Many cases of death of people have remained an unsolvable historical mystery. A terrible monster that claimed many human lives in the years 1764-1767 in Gevaudan of French is usually depicted as a wolf. However according to some eyewitnesses, it moved by jumping on two legs [58] and the sides of it's body were covered with red spots like a Chupacabra in Puerto Rico [37, 43].

In some cases, Chupacabrus even entered a house through open windows and attacked owners on theirs own territory. Echoes of a real presence of this vampire are preserved in the legend about the vampire from Croglin Grange in Cambridgeshire, retold by Augustus Hare in the 19th century. At night, a terrible creature with red eyes attacked Lady Cronwell. The creature, crouching to ground and moving by jumping, approached from the garden side and jumped through the window. This attack was extremely fast. Relatives who came running saved the bitten lady wich wounds on her neck and severe bleeding. The vampire have been hopping away to dark.

Most famous Chupacabrus (Jumping Jack) attacks at human were happened after 1837 around London [40]. He attacked children, women and even men in the London suburbs of Hammersmith, Ealing and Isle, on the country roads near Barnes and Clapham. Many were seriously injured. At that time, many English newspapers, «The Morning Chronicle», «The Times», «The Morning Herald» and other wrote about these events. Housekeeper Mary Stevens described this encounter in the evening near Griffin Park. A terrible red-eyed monster with huge jumps suddenly was in front of her, he tore her clothes with his claws, trying to sink his teeth into her neck. She screamed terribly, people ran to help, but the Devil managed to escape. This was followed by attacks on Lucy Scales, when she walked with her sister along the poorly lit Green Dragon Lane, then on a group of people near a street bar. Gentlemens immediately fled, and monster attacked the bar waitress. Chupacabrus didn't complete the kill, probably someone scared him off. Soon the police found the girl unconscious with heavy bleeding [56, 57, 59]. In 1843 Chupacabrus attacked mail-coach drivers in East of England. Bodies with «deep scratches or burns» were found in hard-to-reach places or on country roads [26, 39, 56].

In 1909, Chupacabrus attacked peasant Ivan Zharkov in the Olemtsy village of Russian Irkutsk province. The creature attacked in the barn inflicting a bite on his neck. The man fought back with difficulty with a shovel [41]. In the same year, the American Chupacabrus (Jersey devil) attacked Mrs. White in her own backyard [65, 66].

Many such incidents have occurred in Latin America. In 1975, in Puerto Rico, Chupacabrus attacked Juan Muniz, who managed to fight back. Twenty years later, Osvaldo Claudio Rosado was attacked the same manner. He also escaped, although deep wounds made by Chupacabrus claws remained on his stomach. The same year in Canovana Chupacabrus attacked Marcos Vigil, his house indoors. Similar incidents are also known in Mexico: Jose Ángel Pulido in Tjomulco and Theodora

Ayala Reis in Sinaloa were attaked (1996). Chupacabrus left claw marks and deep paired wounds on their bodies. Luciano Olimpio di Reis was injured in the city Passos of Brazil (May 1996). A week later, elso three victims became known in Passos [32, 43]. One lady was found dead in her own house. Two of her dogs were also killed. All of them were exsanguinated [89]

Similar incidents happened in the Ukraine. In March 2009 in Chernihiv region Chupacabrus attacked a woman returning from work in the evening. In October 2010, in the Bolgrad of Odessa region, Chupacabrus attacked a man, scratching his face. Several cases of attacks on people occurred in Rivne region. In March 2010, Chupacabrus jumped from a tree onto 14-year-old Victoria Beletskaya. In May 2011, Chupacabrus attacked an elderly woman, Nadezhda Zozolyuk, leaving lacerations on her legs. In August 2011, in village Bobrik of the Sumy region, a smelly Chupacabrus suddenly jumped from a bridge right in front of Ruslan Kosolap, and, having struck him with a sudden blow, knocked him to ground. Car headlights frightened the creature and he got away. Deep scratches remained on Ruslan's hands. In January 2012 night motodepot guard Yakov Gorbachev was attaked in the Artsyz of Odessa region. In April 2016 the same incident was made close of the Volgodonsk of Russian Federation. Chupacabrus was climbed into a chicken coop. Local owner hearing a noise diriged there and was attaked [67-72].

Sometimes Chupacabrus also attacked people moving along the road in bikes. Late at night in the Morris County of New Jersey, Chupacabrus (Jersey Devil) knocked teenager Jim off his bike. In the Ukraine, the same incident occurred in September 2009 in the village Ivanopol of Zhytomyr region. A man returning from work on a bicycle was knocked off his bike, he managed to notice «hoofed legs» only. In the midnight in May 2010 in the village Stepanovka of Vinnytsia region Chupacabrus jumped on biciclyst Maxim Godonyuk. The boy managed to stay in saddle, but there were deep scratches on his back and arms. Cases are even described when Chupacabrus tried to attack passengers of parked cars (Jersey Devil and Mothman) and even a tractor driver (Chupacabra) (the Penza region of Russian Federation, 2018) [62, 73-76].

Perhaps an adult man is quite capable of fighting off a monster, especially if he does not lose his head and is armed with at least a stick. However, Chupacabrus pose a serious danger to children who have no chance to protect themselves from the monster. On the night of October 27, 1967, in the Rolling Prairie of Indiana, a Chupacabrus broke through a leaky roof of a family trailer and attempted to steal a child [77]. A similar incident occurred in 1970 in the Mercer County of New Jersey, where a Chupacabrus (Jersey Devil) grabbed a child by the hair [62]. On March 10, 2017 the YOUTUBE «DENISOVGAMES channel» showed the story «The werewolf in Kursk region». A strange creature very similar to a Chupacabrus was chasing schoolchildren returning from school through the forest. It tried to attack one of the boys. This event took place in the mid-1990s.

However, the most terrible deaths of such attacks, the security authorities try to classify as much as possible. On story was investigated by Eugenio Baamonde, and later mentioned by Scott Corralles in the article «The sum of all fears» in 2007. In the late 1970s, 26-year-old Juan Miguel Santos and his fiancée died near the Punta Arenas of Chile. The death of a young couple occurred in a country house where they went to find out who was killing and bleeding their rabbits. It was reported that death was accompanied by strange circumstances, however, according to the Chilean researcher Raul Nunez from the IIE organization, this case was classified.

The tragedy that occurred in 2011 in the village Berezovka of Khmelnitsky region of the Ukraine, local authorities tried to present as an attack by a rabid dog. There were several bitten victims. The worst incedent event was happened to 21-year-old Veronika Sugak, who later died from wounds and neuroinfection. The attack happened in the morning at a rural bus stop. A local women resident who witnessed the attack reported that «it was a monster the likes of which she had never seen in her life.» It pounced on the girl, tearing off and immediately devouring pieces of her body: a piece of her leg, fingers and lip. He tore off part of the scalp. Attacking, the creature stood on two legs, and its height reached two meters. The husband of the witness with a gun came running to the screams. Being an experienced hunter, he nevertheless shot the monster many times before he was able to kill him. The Chupacabras corpse was hastily taken away and supposedly cremated. The details of this case are classified [78, 79, 94].

But there have been worse cases. In 1962 in Maryland a Chupacabrus known as «Goatman» killed 14 travelers. He literally tore people apart. The horrifying circumstances of their death are classified. In 1999, a geological expedition went missing in the Potosi province of Bolivia. They were found dead. Almost all of the victims died from wounds inflicted by the claws. The investigation found that all five geologists died at night. No one had time to call by satellite phone or by radio. People tried to fire pump-action shotguns. Shooting was carried out chaotically, in all directions, sometimes upwards. The killer attacked at the moment of reloading the weapon, methodically killing one person after another. The last man was killed when he ran out of ammo. He tried to run away, but the predator caught up with him in a jump. Two round holes were found on his neck [80, 81].

In 1996, in India, in the state Uttar Pradesh of India Chupacabrus killed two dozen children. In the same state, in April-May 2001, many people were injured after. People who were sleeping on the roofs of their houses on hot and stuffy nights were attacked. In 2002, attacks on residents of the state were repeated again. There were many wounded and even seven dead caused by wounds and burns. Ragurai Pal the resident of the city Shanwa of Gujarat died of wounds, his stomach was torn. In November 2017, a 10-year-old boy with a bite mark on his neck died in Tikaria village of Sitapur. In January 2018, another eleven child deaths were happened in Hyderabad. On May 13, 2018, near Sitapur, Chupacabrus attacked a 12-year-old girl [83 – 88].

Not all deaths from Chupacabrus bites and their claws have been properly investigated. Suspicious deaths have occurred, for example, in Honduras, in the Usinsk region of Kamchatka, and the Knott County of Kentucky [90 – 93]. Usually investigators were unable to determine what kind of predators attacked people.

Chupacabrus seem to be very intelligent animals. Their insectoid nature is beyond doubt. However, at the place where the Chupacabrus were shot, dog tracks and even the corpses of strange hairless dogs are sometimes found. This is strange because any dog can not inflict symmetrical paired wounds 12 cm deep and completely suck out the blood of a victim. Any dog can not also tear and destroy metal barriers and cages. So it can be assumed dog corpses were planted by security agents in an attempt to disguise a nature of real Chupacabrus vampires. On the other hand, «dogs» could accompany Chupacabrus and are «expendable material» in case of shooting [95].

MEETINGS with UPPER ALIEN CASTES

Encounters with other castes of Alien Insectoids outside of underground isolates occur not too often, less frequently than with Chupacabras foragers. It was mentioned in REPORT-96 that members of Upper castes suffer from oxygen poisoning after several hours of being on the surface. However, if these meetings had happened after all, then under circumstances that clearly indicated their reasonable behavior, possession of objective activities, a certain level of knowledge and high technology. Sometimes these creatures arrived and departed on UFOs, thay supervised the medical examination of an abductee and performed some subtle manipulation, or shared with contactees some knowledge about the world around them. As a rule, they have huge eyes and have not grasping near-mouth organs. Their head appears triangular due to the large size of their eyes. Therefore, they are often compared to huge praying mantises. In addition to the High castes, there are other castes that have smaller eyes.

M. D. Graeber, the founder of the Philadelphia UFO Center, wrote about Alien Insectoids. The well-known ufologist D. Carpenter has collected a whole archive, where there are about twenty testimonies of meetings with them [1].

In August 1945, in the vicinity of San Antonio in New Mexico, local boys Reme Baca and Jose Padilla had saw a fallen UFO, they told about this in the program Coast-To-Coast-AM. Padilla recalled a sonic boom similar to an atomic bomb test.

According to him, the crashed object had a diameter about 10 meters. The boys looked into the portholes of the ship, where one of them noticed a huge «praying mantis» [2].

In 1963 Linda Porter, 17 year old resident of Covina in California (1963), was abducted from her bedroom. She got in some kind of isolate with gloomy corridors. There among other creatures she saw a «Giant Praying Mantis». He accompanied Linda and talked to her. This event was the beginning of many years of contacts, in which Linda was allegedly recovered from a heart defect and received knowledge about the «real structure of space and matter» [3]. In 1972, Judy Kendall, also a resident of California, saw the same creature [5].

Giant Praying Mantis by Linda Porter (1963)

Insectoid portrayed by Simon Parkers

In 1972, Judy Kendall, also a resident of California, saw the same creature. В 1976 г. Jack, a resident of Maine wached strange ugly creatures with huge compound eyes without eyelids and with «hands» resembling insect paws with four fingers [4]. In 1967 a mantis-like «doctor» treated 22-year-old Lewis Whitley Strieber from San Antonio of Texas (later Striber became a well-known researcher of these phenomena) [6]. Simon Parks, an English politician from the Labor Party, portrayed such creature that he saw in childhood when was subjected to abductions.

The memories of some huge mantis-like creature have been preserved from childhood and by Carla Turner (since about 1952). This creature forced her to obey, holding her by the shoulder. The girl was crying and screaming that it was not her mother. Eight-year-old David Huggins stumbled upon a huge Praying Mantis in the yard of a farmhouse in Georgia, where he lived with his parents (1950). He later painted this creature. Abductee Jean Robinson also reported similar contacts [6, 7].

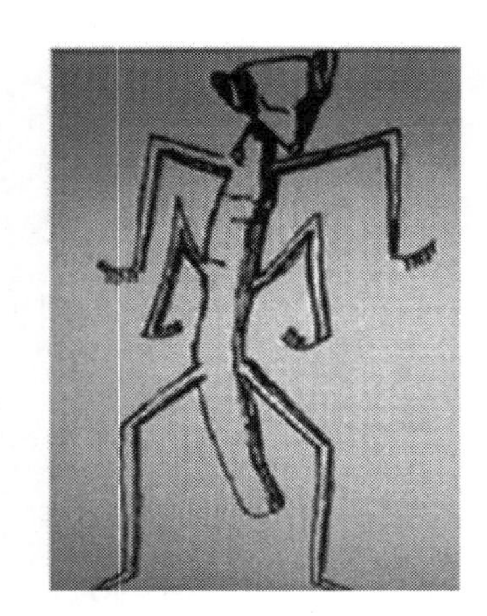

Giant Praying Mantis by David Huggins (left)
and by Carla Turner (right)

Other contactees described creatures with a «crest on their heads». Joshua Rhinehall, a private researcher from the Pacific Northwest, also described his encounter with an insectoid. Its head was approximately 50 cm long and 33 cm wide. At the top of the head was a raised wide flat crest. The eyes were a solid black shiny surface. The wrinkled dark gray skin on the face had a different texture. The large mouth opening resembled a slit. The neck was very thin and looked like a finned tube. The creature was dressed in a tight gray suit, the arms were extremely thin [19].

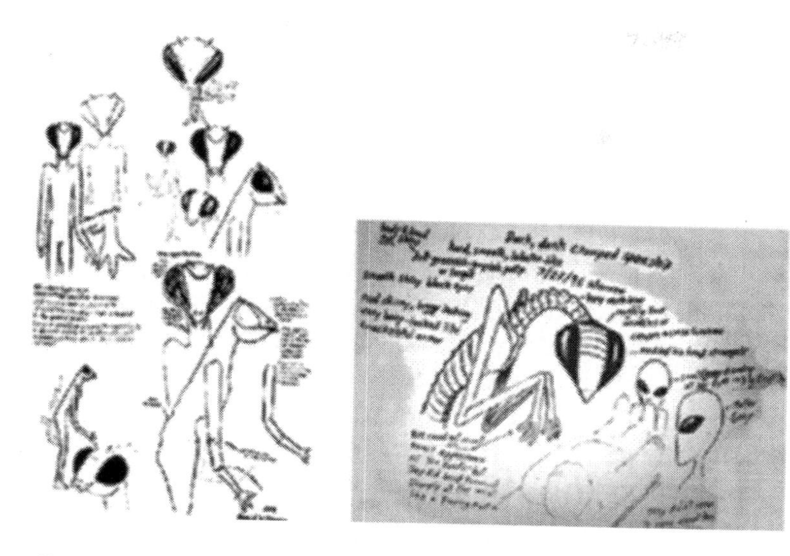

Insectoids by David Chase (1995) and Jeanett Roussel (1997)

Insectoid by Joshua Rinehall

In 1980, in the area of Cambridge in the Canadian province of Ontario, a lady was abducted along with her child. Then she remembered being led through a long tunnel with jagged edges. She was accompanied by creatures that looked like huge insects (1980) [11]. In the same year a resident of New Mexico, Mirna Hansen, was kidnapped right from the car along with her 6-year-old son. Some strange non-human beings brought her to their base or nest, presumably near Dulsa. She has been forcibly subjected to medical procedures. A team of «medics» was led by a tall creature in a white cloak with a hood. His huge eyes, half the size of his face, were lidless, his face was «bony, structured.» The creature had long fingers and huge claws, they burned, a burn remained on the woman's forehead [8]. Entertaining portraits of «praying mantis» aliens were created by David Chase on the basis of such reminiscences (1995) and Jeanett Roussel (1997) [10].

The most impressive description of the insectoid was left by David Turner (a pseudonym). He was abducted in October 1988 in Oklahoma and also subjected to inhumane medical experiments. Waking up after losing consciousness, he realized that he was lying on a metal table, and around him were several insectoids. He managed to see one of them from a very close distance. It was a tall (about 180 cm) massive creature resembling a praying mantis. Some clawed outgrowths were on his head, soft folds of skin, like rag or scarf (lung sacs) were on his neck near his head. The eyes were large, round and dark, with an elongated and pointed mouth in front. It had long «arms» resembling «broom handles» (pipes, 3-4 cm thick). The creature's hands, flat and wide, were like cancer claws, with two fingers, long, fixed and a smaller, movable one that came out of the wrist. They were sharpened like claws. The «hand» was a little sticky and, as if, in a mitten of their snake skin. The body of the insectoid appeared to be smooth and skinny, like a pencil or a cylinder, in the middle region about 30 cm wide, it tapered towards the neck. He seemed hunched over. He wore what looked like a wide, hard, orange-brown belt. The creature deftly controlled the equipment [7].

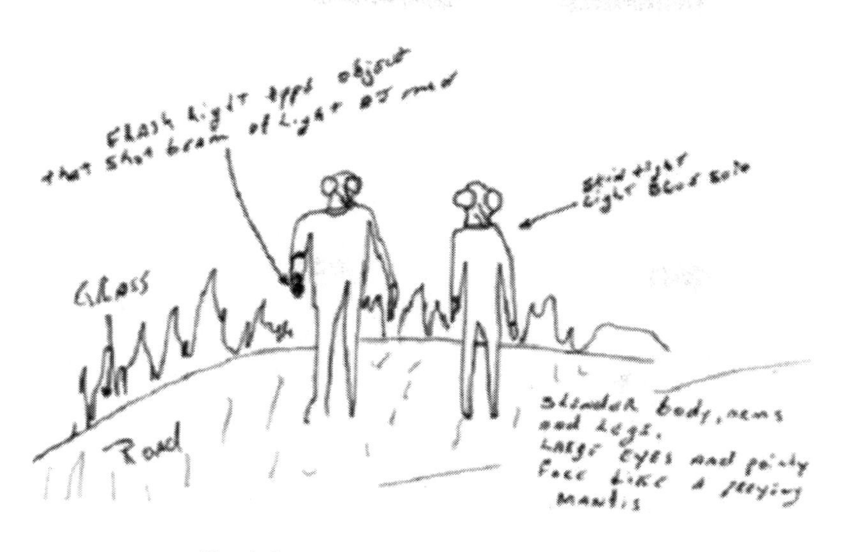

Kevin's grasshoppers. Yukon, Canada

One morning 1987, Kevin, the resident of Canada's northern Yukon Territories, noticed a flying UFO, when he were riding a motorbike down the road. The UFO disappeared around the corner, and Kevin stopped the motorcycle and went around the bend to observe this phenomenon. Suddently, he stumbled upon two non-human monsters, which seemed to him like huge grasshoppers dressed in blue tracksuits (1987). After this meeting, Kevin had a blackout [13].

Many observations have been made outside of abduction situations. Cryptologist and writer Brent Swanser described several real encounters with giant mantise cryptids in New Jersey, Connecticut and California [9]. Some of the evidence is interesting because the observations were made not during abductions, but against the background of natural landscapes, near rivers and swamps. In one case, a fisherman observed a tall mantis-like creature in a «black cloak» that strode straight across the shallow water of a river.

There is one latest event. On July 16, 2020, 26-year-old Paul Froggatt was returning home from a night shift in a dog food factory warehouse. He rode his bicycle home through the woods in Warwick. Driving through Oakwood Preserve and Blacklow Spinney, he suddenly saw a glowing orange sphere hovering over the horizon in the sky. Paul stopped and took a few pictures. At that moment, the object began to move and, as it seemed to the guy, to pursue him. Soon the cyclist drove into the forest and lost sight of the object. At that moment, a strange dead silence reigned in the forest. Suddenly, Paul saw a huge two-meter praying mantis. He stood on two legs, was light green, with a triangular head and large oval black eyes. Paul was immobilized. The boy and the praying mantis looked into each other's eyes for several seconds. It seemed to Paul that this creature was evil and read his thoughts. Perhaps it was preparing to jump. The guy, shaking off his numbness, rushed to run. Later he painted this creature [12].

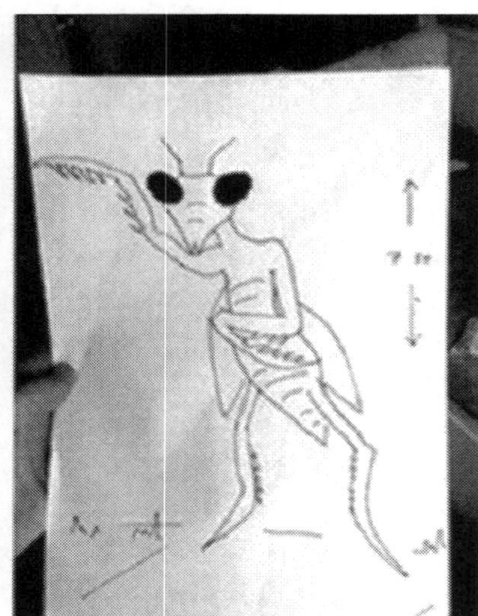

Paul Frogatt and his Praying Mantis

In 2021 Igor Mochalov, a researcher of the Kola Peninsula anomalies, told about a strange huge «spider, sitting on the stone» . This creature was large, about the size of a calf. There was something clearly «mechanical» in him. Navy sailors also observed this strange creature from ship deck. This was a giant spider or insectoid walking along the shore near the Kola Peninsula [14].

On the head of some insectoids contactees saw processes of various shapes. Witnesses may have seen the feathery antennules described in REPORT-96. Differences in the head processes of insectoids are probably due to caste, as well as acquired genetic differences. In addition, it also deals with the practice of the so-called «prenatal surgery» accepted in this strange community. Giant insects with some kind of «hands» or tentacles on their heads seem even stranger. Perhaps, «hands on the head» are called antennas growing near the lophophore.

The offshoots of some creatures have been compared to *combs.* Horned heads seemed to other eyewitnesses. The case mentioned above when getting lost in the caves under the Zhiguli mounts, an amateur speleologist from Samara stumbled upon something like an incubator or a columbarium. After long wanderings in the dark along the half-filled passages of the cave, he discovered clearly artificial, dimly lit corridors with smooth walls. Along the walls in boxes with transparent walls, similar to ice blocks, some two-legged creatures with huge compound bulging eyes «stood» motionless. They had small crooked paws pressed to their stomachs. The lower part of the torso was «rolled into a tube and pressed against the stomach.» The proportions of the motionless creatures did not resemble human at all. The monsters were like two drops of water similar to each other, although some had «bumps» on their heads at different stages of growth (from small to large) [18]. Probably, the «bumps» were a growing lophophore. However head bumps were absent in some creatures, such a head formation may not be characteristic of all castes.

Four-year-old Jane, abducted from her grandmother's house in San Antonio, Texas, ended up in the medical laboratory of some isolate (circa 1947). She was examined, and later taken to a beautiful room with plants and a huge pool. There, a green creature was swimming in the water with luminous «white» eyes, a long neck and some kind of «handles» on a large head. The monster grabbed the girl by the ear, she screamed terribly. She didn't remember anything else [21].

Juvenile Casey Turner, abducted in Kansas from his father's car, was taken somewhere where he underwent a medical examination. He remembered a tall non-human creature with black eyes, a crest on its head and vertical furrows on its «face» (early 1960s) [7]. A tall creature with the same crests visited Ted Rice on numerous occasions while he was living in Atlanta, Georgia (late 1960s). Its body was gray or olive brown, and its bulging eyes without lids or pupils were yellow or golden in color. In 1980s members of Karla Turner group saw how insectoids with «combs» were involved in human kidnappings in New Jersey [20, 21].

In the summer of 1951, the city of Belgorod-Dnestrovsky, Odessa region, was seized by panic. People talked about strange «horned» creatures appearing at night from the dungeons of the old Akkerman fortress. First they were seen by four teenagers fishing at dusk. A few days later they were noticed by a pensioner who went out at night to smoke. The police received a statement from the driver, who was attacked in the dark by stinking horned upright creatures, about human height. The sorties of these creatures continued until the mid-1980s. One night, local residents heard underground explosions, after which the night visitors did not appear [22].

In some cases, eyewitnesses did not even think of somehow connecting these very strange creatures with insects. This is understandable because appearance of a «upright insect» of such an unimaginable size is too unusual for us (sometimes it could be named «reptiloide»). Some insectoids are seen wearing clothes that almost completely hide their bodies. Sometimes they wear special breathing masks. The fact that they still belong to insectoids can only be judged by certain signs, for example, these are compound large glowing eyes without pupils and eyelids, claw-shaped limbs, legs «growing from the waist», or legs with «knees back», etc. The close presence of an insectoid creature gives off an odor of hydrogen sulfide or mercaptans.

In 1952, Frank Fescino, Frieddly May, and several other witnesses in the Flatwood of West Virginia, observed a creature with bright red eyes. It emitted a terrifying odor that even caused some observers to faint. Apparently, the creature arrived on a fireball (the ball was at a distance) [15].

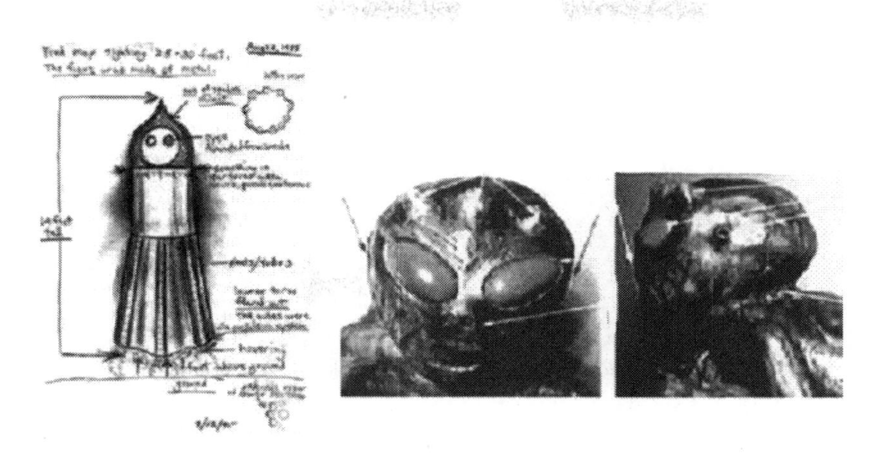

Red-eyed monsters from Flatwood (1952)
and from Varginha (1996)

Several such meetings took place on the territory of the former USSR. Andrey Rogov, one of leader of Institute of Robotics and Technical Cybernetics in St-Peterbourg, spoke about the events that happened to him in the 80s of the last century. While on expeditions, in the dungeons of the New Jerusalem Monastery in the Istra of Moscow Region, he stumbled upon a strange straight creature, about the height of a man. He had glowing bright red eyes with no pupils. He was clearly sentient and was wearing a hooded cloak [16]. The same monsters were observed in the 1970s in the Leningrad and Donetsk regions and in 1990s two such monsters visited Irina Yampolskaya in the Voronezh region.

To insectoids, apparently, one can also include those monsters that were caught by firefighters and military gardes in Braziliain Varginha (January 1996). This happened within the city, so many people saw these monsters. Three girls examined one of them especially well. It was something with red bulging eyes without pupils and brown oily integuments of the body. From him and his bird tracks came the suffocating smell of mercaptans and ammonia. One of the monsters, being surrounded, produced a buzzing sound similar to the buzzing of a bumblebee [17]. According to eyewitnesses, a reconstruction of the appearance of the monster was made. Judging by the size and external soft integument, they could be larvae after the imaginal molt.

At the end of June 1979, children who were resting in the Beryozka pioneer camp near Derzhavinsk, Turgay region, Kazakh SSR, ran away from tall, upright creatures with an absurd gait [23]. They moved with their legs high, like soldiers on the march, with long «arms» stretched forward. Their legs grew «from the waist.» The creatures were black and thin, with some sort of skirt dangling around their hips (It may have been a belt similar to the one described by David Turner – see above). No noses or mouths were visible on the faces, only two large «eyes» of pink color. And one of the camp mentors at night came across the same creature with red eyes and a long «horse» muzzle. Approximately the same creatures were seen by teenagers ten years later in village Konantsevo of Kharovsky district in Vologda region [24].

Most likely, insectoids were observed by vacationers at a sanatorium in Blankya, near the Sofia of Bulgaria in 1989. The growth of strange creatures was about three meters, they were dressed in overalls, and their faces were hidden under masks. They had long «arms» and walking they raised high their legs. Their legs their legs grew from the waist [25].

Thus, it can be argued that insectoid creatures, although they lead a secretive lifestyle, meet people more than once. Some of these meetings are reflected in tolking of dumbfounded eyewitnesses. However, these stories are not taken seriously.

NOTES to Part 3

1. TV «Coast to Coast AM», 25 May 2017 – «Aliens & UFO Encounters»

2. ufoevidence.org/case852

3. Howe LM «Glimpses of Other Realities», V2, 1998

4. Fowler RE «The Allagash Abductions», 2005

5. Hall RH «Judy Kendall abduction, three types of beings»// «The UFO Evidence: A Thirty-Year Report», Scarecrow Press, 2001, 704

6. Strieber LW «Communion», 1987

7. Turner K «Into the fringe», 1992

8. Howe L «An Alien Harvest. Further Evidence Linking Animal Mutilations and Human Abductions to Alien Life Forms», 1989

9. Brent Swancer «Truly Outlandish Encounters with Insectoid Mantis Men», October 30, 2019

10. Chace DW «A visual guide to alien beings», 1996

11. Wheeler B «Cambridge ON Contact Report» – 18.11.1980 // Cambridge UFO Research Group, Ontario, Canada www.pararesearchers.org/index.php/other-world-entities/489-cambridge-on

12. Giant praying mantis attacks UK warehouse worker after taking UFO pictures – КРАТКО-НЬЮС 31-01-2021 – kratko-news.com/2021/01/31/gigantskij-bogomol-napal-na-rabochego-sklada-v-velikobritanii-posle-sdelannyx-snimkov-nlo/»

13. www.ufoevidence.org/cases/case100.htm

14. Loki Wotan «Tales of Grandfather Shaman. Evil Spider» – www.youtube.com/watch?v=x4bqsDaYgmA&list=PLGWsC6thRojkuZ6jVBhXmSM8A5wS3GXQn&index=6

15. Gershtein M, Deruzhinsky V «Undead or Mysterious Creatures», 2015 (In Russian)

16. Film RenTV, «The Secret of Supercivilization. Underground Cities», 2015

17. Pacaccini V, Ports M «Incidente em Varginha: criaturas do espaço no Sul de Minas», 1996

18. Pavlovich IL, Ratnik OV «Secrets and legends of the Volga dungeons», 2003 (in Russian)

19. From the correspondence of abductees with Linda Howe: Subject: Ant heads, Date: January 26, 2011 To: earthfiles@earthfiles.com – freerepublic.com/focus/chat/2597204/posts?q=1&;page=156

20. Turner K, Rice T «Masquerade of Angels», 1994

21. Turner K «Taken: Inside the Alien-Human Agenda», 1994

22. Lyakina E , SECRETS OF THE USSR, June 2021, No. 10
23. EXPRESS K, 20.04.2013
24. «Kharovskaya landing»// Around the world, 1989, 10
25. Azhazha V «Another Life», 1981
26. Fort CH «Lo!», HYPERTEXT EDITION BY MR. X – www.resologist.net/
27. Chernobrov V «The Mystery of the Chupacabra», report 09/27/2016
28. TV Programa «Dimensiones Paralelas», 20 – Analizamos el caso de Orlando, 9.02.2019
29. Diario Austral de Temuco, 18 Feb 2001
30. RUSSIAN UFOLOGICAL DIGEST, 9.02.2001
31. Voronova N «Chupacabra monster of the Chilean desert» – ABNORMAL NEWS FROM AROUND THE WORLD – ufonews.su/text2/568.htm
32. Nepomniachtchi N «Chupacabras: unknown in nature or space vampires?» – AROUND THE WORLD, 1999, 433. «GAZETA.UA» 10.02.2010 – gazeta.ua
34. «With the onset of autumn and cold weather, the Chupacabra did not calm down» – PARANORMAL NEWS – 24.10.2011 – paranormal-news.ru
35. FACTS (Chernihiv – Kyiv),7.10.2010 (in Russian)
36. «SOBESEDNIK» (Moskow) -11.10. 2001 – N 40, p 9 (in Russian)
37. Redfern N «Rofd Trip» Llewellyn Publications, 2015, 264 p.
38. Agon E, Shevik S, Alkor E, «The Mysterious Devil Chupacabras», 2019 (in Russian)
39. Dash M «The Devil's Hoofmarks source material on the great Devon mystery of 1855», FORTEAN STUDIES ACADEMIA.EDU, 1994, 71-150
40. Gershtein M «SECRETS OF THE XX CENTURY», 2010, 21(in Russian)
41. slavyanskaya-kultura.ru/Slavic/gods/vstrechi-s-nechistyu.html
42. Newspaper «RADOMYSHLYANIN»- 6.02.1913 (in Russian)

43. Eberhart GM, «Mysterious Creatures: A Guide to Cryptozoology», 2002

44. Gaiduchik In «On the Threshold of the Unknown» УФОКОМ, 4085 – www.ufo-com.net/

45. Diario Austral de Temuco, 18 Feb 2001

46. «The Stafford werewolves have been haunting the locals for six years now» -hronika.info/neverojatnoe/1666-staffordskie-oborotni-uzhe-shest-let-ne-dayut-pokoya-mestnym-zhitelyam.html

47. UTRO.RU 04.05.2010 – www.utro.ru/articles/2010/04/05/885667.shtml

48. «In the south of Spain there was a meeting with an animal resembling a gargoyle», KRIPTOZOO.RU 30.11.2017 – kriptozoo.ru/

49. Valle J «Passport to Magonia: on UFOs, folklore, and parallel world», 1993

50. Thompson W, «Houston batman», Oct 1953

51. Zuev V «Aral dead end», 1991 (in Russian)

52. «The connection between the appearance of UFOs and Chupacabra was found in the Ukraine», UKRAINE ANOMAL 19.08.2011

53. «What kind of monster?», A plot from the TV cycle «We Speak and Show», NTV 08/5/2013 – www.youtube.com/watch?v=fMpjdpxq4LM

54. Azhazha V, Zabelyshensky V «UFO. Contacts», 2009 (in Russian)

55. «INDIA TODAY» 17.05.2018

56. Dash M, «Spring-heeled Jack: To Victorian bugaboo from Suburban Ghost», Fortean Studies, 1996, 4, 1-125

57. Begg P «The Terror of London», 1981

58. «Beast of Gevaudan» - www.unknownexplorers.com/ beastofgevaudan.php

59. Haining P «The legend and crimes of Spring-heeled Jack», 1977

60. Villiers E «Stand & Deliver», London, 1928

61. Downes J, Davies G «Owlman and Others», 1997

62. Leuter K, «The Devil Hunters» – OFFICIAL RESEARCHERS OF THE JERSEY – www.njdevilhunters.com/art19251214.html

63. Darcy Weir «The Underground: Director's Cut», 2020

64. Anfalov AA «Aliens in the Crimea», 2018 (in Russian)

65. Moran M, Sceurman M, «Weird N.J.: Your Travel Guide to New Jersey»s Local Legends and Best Kept Secrets», BARNES & NOBLE, 2004

66. «In 1909, the Jersey Devil was sighted in Gloucester», CNBNewsnet, Feb 2007, www.gloucestercitynews.net

67. FAKTI (Ukraine) 26.03.2010 (in Russan)

68. FAKTI (Ukraine) 3.11.2010 (in Russan)

69. Bychkovskaya L «In the Rivne region «chupacabra» attacked an 80-year-old woman», FAKTI (Rovno) 12.05.2011

70. KOMSOMOLSKAYA PRAVDA in the Ukraine, 13.08.2011

71. «A Chupacabra attacked a resident of the Odessa region» PARANORMAL NEWS 01/30/2012 – paranormal-news.ru/

72. «Chupacabra attacked a resident of the farm Maryina Roshcha» 3.04. 2016 – www.youtube.com/watch?v=GKx1BsOhQhc

73. «Lviv Chupacabra. Photo and video report», UKRAINE ANOMAL 24.06.2009 – ufodos.org.ua/

74. Sergushev M, FACTI (Vinnitsa-Kyiv) 14.05.2010 (in Russian)

75. Bykov About «Chupakabra: in the Kuzbass village of Bekovo the horror settled?» 7.02.2019 - avoka.do/posts/chupakabra-v-kuzbasskom-sele-bekovo-obosnovalsya-uzhas

76. «The man told how in his childhood he almost took a flying humanoid», CRYPTOZOOLOGY NEWS 5.03.2019 cryptozoologynews.com

77. «The man told how in his childhood he almost took a flying humanoid», CRYPTOZOOLOGY NEWS 5.03.2019 cryptozoologynews.com

78. GAZETA.UA 12.08.2011 (in Russian)

79. GAZETA.UA 11.04.2012 (in Russian)

80. Couch JN «Goatman: Flesh or Folklore?», 2014 – www.jnathancouch.com

81. Project «PREQUEL» – chronology – prequel.rf/chronology.htm83.. «Spring-heeled Jack in India», FORTEAN TIMES, 1996, 91, October, 20

84. «Fears of a Spring-heeled Jack in India», FORTEAN TIMES, 1996, 92, November, 15

85. Crossbows A «Indian monster» – 4stor.ru/histori-for-life/3521-indijskij-monstr.html

86. Nepomniachtchi N «100 great secrets of the East», 2008

87. «Aggressive Indian Invisibles» PARANORMAL NEWS 12.09. 2015 – paranormal-news.ru

88. INDIA TODAY, 17.05.2018

89. Shlionska I, «Encyclopedia of evil spirits», 2006 (in Russian)

90. Gisselle Flores «Siguatepeque: el «Chupacabrus» -LE TIEMPODIGINAL 02/13/2019-tiempo.hn/Chupacabrus-en-siguatepeque-mata-a-jornalero/

91. «A shift worker in Komi was bitten to death by an animal», RU NEWS – 15.10.2020

92. vz.ru/news/2020/11/4/1068890.html/

93. FRANCES MULRANEY – DAILYMAIL.COM-17.10.2020- www.dailymail.co.uk/news/article-8024859/Boy-13-killed-animal-attack-officials-dont-know-creature-killed-him.html

94. It should be noted that Chupacabrus is generally quite difficult to kill, given the system of regulation of the vascular lumen inherent in insectoids. Vessels at the site of injury immediately constrict, isolating that area of the body from blood flow. However, sometimes Chupacabrus manage to shoot. Once, blue spots remained at the place where the Chupacabrus was at the time of the shot. Alien blood is blue, stained with the copper-containing pigment hemocyanin.

95. There is one very strange circumstance. These dogs are not only hairless, they completely lack hair follicles like the Grays, the nature of which will be discussed below. It is possible that both, Gray and hairless dogs, are the product of Aliens selection carried out in the same direction. This interesting hypothesis was expressed by our colleague Sergey Shevik, but it needs to be verified.

PART 4.
ANTHROP SYMBIONTE SLAVES

APPEARANCE and PHYSIOLOGY

Traditionally, the so-called «humanoids» or «EBE» («extraterrestrial biological entity») are considered the main type of «extraterrestrial aliens». The American special services have made a lot of efforts to make them popular, and on other hand, to sow doubts about their real existence. It seems suspicious elso that last time the image of humanoid aliens is being strongly imposed on society as «declassified archival documents» of American intelligence services [1, 2]. As a result, completely unjustified meaningless excitement rises around this problem.

In this topic, everything is false, even the terminology. First, the term «humanoid» carries the wrong semantic load, because they used to mean an extraterrestrial creature, like an artificial «biorobot», which has only an external resemblance to human. Secondly, the term EBE, adopted in the 1940s, has also become obsolete, as evidence of an extraterrestrial origin for these strange creatures has never been found. Therefore, in the future we will call these creatures «Anthropoids» or simply «Anthrops», implying that these representatives of an alien civilization have not only an external resemblance to humans, but also common earthly ancestors with them. According to modern anthropology, the subfamily Homininae includes gorillas, chimpanzees and humans and Homo sapiens sapiens only is surviving human representative.

Finally, neither the Aliens nor the Anthrops are anything extraterrestrial. The Anthrops were not the main specie in the socium of Aliens, but *just a symbiotic species* in relation to Insectoids that were living with them in underground isolates.

Like Aliens, Anthrops are products of natural biological evolution. However, the formation of their breeds occurred by artificial selection. There are a many such domestic breeds: Greys, Nordics, Gnomes and so on, are obviously a *result domestication of wild human species* of our human predecessors. And we will try to prove it below.

As a rule, the bodies found in crashed flying objects are mainly of the Gray Anthrop variety. In underground isolates elso, they swarm like an ant, doing all the domestic chores. Most likely, Anthrop symbionts live in spatial underground isolates of Aliens in the position of domesticated animals. James, who was kidnapped in the middle of the night, got in Alien termite mound, an underground nest of aliens. He saw a lot of the same little «gray» creatures busily scurrying back and forth in different directions with small flashlights on their heads [3, 4].

Grays are an artificially race breed of Anthrops that are adapted to breathing in of low oxygen atmosphare. This quality is very useful in stuffy underground rooms, so Grays often wear breathing masks. Insectoid Aliens treat this kind of Anthropoid slaves no better than they treat expendables. According to abductees, Grays perform most of a practical work inside and outside a nest. The Grays were seen more than once by members of Carla Turner support group. Grais take soil, water or animal blood samples to analyse some sort of device [3, 4, 5]. They also hunt by flying over fields and pastures in UFOs (see below).

Existence of Anthrops is quite real. However along with real images, many fake portraits were published, sometimes even toys from well-known stores. Some published videos contain obvious animation elements. Nevertheless, despite this, the real existence of humanoid aliens is beyond doubt and we cannot completely discard the testimonies of numerous witnesses and we can only make an attempt to evaluate the information about them that came from various sources. Among the large number of Anthropoids, little people, the so-called Grays, were the most common.

Anthrop pilotes or their bodies were found in UFO crash sites, information about them was painstakingly collected in the multi-volume current reports [6]. Turns out UFOs were crashed many times over different parts of the world, especially after US

Air Force began to use laser weapons against them. Similar catastrophes happened in particular, the summer of 1947 in New Mexico and one in January 1950 in northern Mexico, and at least, two crashes in the USSR (Russia) [7]. Большое количество аварий однозначно свидетельствует против внеземного происхождения социума Чужих.

According to different sources captured UFOs and their pilots studies were carried out in complete secrecy in many American research centers. The most frequently were mentioned Indian Springs Air Force Base (now Creech), the Area 51 (other code names were Dreamland, Paradise Ranch, Home Base, Watertown Strip, Groom Lake, Home) of Nevada, the Sandia Lab of Albuquerque (New Mexico), the NASA Langley Research Center of Hampton (Virginia), and the US Naval Headquarters Indian Unit of Maryland. The surviving captured Anthrops were allegedly examined in multi-level complexes equipped with special chambers with adjustable atmosphere composition and pressure.

Robert Lazar allegedly saw photographs of autopsy of Gray Anthrop body and even a survived Anthrop (although he was not sure about this) [8]. The presence of Grey Anthrops at military bases was confirmed by USMC Air Force Captain Bill Youhouse, as well as a soldier who wished to remain anonymous, who for 12 years was responsible for logistics inside Area 51. Derek Henessy, a former «Dreamland» guard of underground complex sow Grey's bodies in a conservation solution [9]. Former Lockheed Martin Senior Scientist Boyd Bushman also claimed that there were Anthrops at the bases, although he presented dubious photographs as evidence.

Retired Air Force Lieutenant Colonel Philip Corso (1915–1998) had served in the intelligence service during the administration of President Eisenhower. At the end of the last century Corso published a book, «The Day after Roswell», which made a lot of noise [10]. According to Corso, military reports mentioned various locations near Roswell, in San Agustin, and in Corona. Corso was not a direct witness to the crash. The description of one incident was made by Intelligence Major Steve Arnold. In addition, there were discrepancies in Corso's book regarding the date and place of the incidents. However, Corso must had some credible information because have being Chief of the Pentagon's Foreign Technology desk in Army Research and Development (working under Lt. Gen. Arthur Trudeau) he had access to classified documents.

Another information source about strange creatures was William (Bill) Milton Cooper (1943 – 2001) [11]. After World War II, Cooper served in the intelligence department of the headquarters of the US Pacific Fleet, where he had the opportunity to read secret papers. There, he allegedly saw photographs of dead UFO pilots. Many ufologists considered Cooper a charlatan. However, in 1977, former Air Force officer William English, in an interview with Linda Howe confirmed the existence of such photographs. He allegedly saw their color version in Report No. 13 of the US Air Force Project «Grudge». Some information (or disinformation) about Anthrops was translated by special agent sergeant Richard Doughty and became known from a short report intended for US President Dwight D. Eisenhower [5, 12].

Contactees have also seen Anthropoids on numerous occasions. As a rule, memories of them surfaced under hypnotic regression. Often, regression hypnosis was used to restore memories of strange «adventures». However, some contactees had quite conscious memories. Portraits of these creatures can be seen in the books of Whitley Strieber, Budd Hopkins and Linda Moulton Howe. They were remembered by patients of American psychiatrist, Harvard professor, Dr. John Mack (1929-2004), and members of support abducted group, that was formed by Carla Turner (1948-1994). A small video was also shown as a leak from their Russian State Security Committee («KGB») [13].

We have no way of verifying the truth of statements from different sources. However, one should compare them with each other and try to determine whether they can correspond to any reality at all. And, if so, could these beings have some «extraterrestrial» features.

Philip Corso recalled how General Twining showed him photographs of «EBE» and the autopsy results of their bodies. *The medical experts, who are in captivity of the accepted extraterrestrial paradigm, were most puzzled by similarities of these creatures with our biological species* [10].

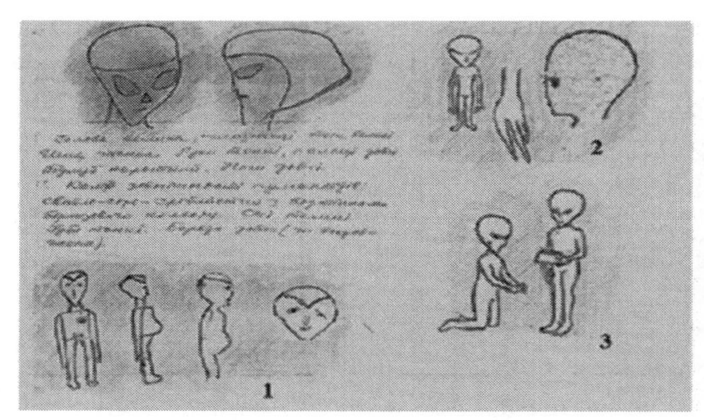

Grey's portraits provided by Vladimir Litovka (1):
(«Big head, big eye sockets. The neck is thin, the arms are thin, the fingers are long. The body is short, the legs are long. The skin color is light silver with a turquoise tint. The eyes are dark, the lips are thin, the chin is long or elongated») and from books of L. Springfield (2) and K. Turner (3), Skinny Bob from «KGB» (4)

Grays were different sizes: 80 – 150 cm. However there are no significant contradictions between descriptions of appearance and portraits made by eyewitnesses, although their size, head shape can vary between they, but no more than in Homo sapiens populations. According to information received from Doughty, Cooper, Corso, as well as abductants, these creatures were very small, their body length was approximately 130 cm, and weight – 18 kg. There are no significant contradictions between descriptions of appearance and portraits made by eyewitnesses, although their size.

The contactee, whose testimonies and drawings were kindly provided to us by the Ukrainian ufologist Vladimir Litovka. The same portraits of these creatures are present in the books of Springfield and Turner. According to eyewitness accounts, they are sentient, emotions show on their faces and ables communicate using «telepathy». Contactee Amy sketched the same creatures after visiting a military base. In her picture, they are sitting at tables using human furniture. However, they are one and a half times smaller than a person of average height. Another abductee, Angie, portrayed two little people, one kneeling, the other holding an object in his hand. They have completely human hands and feet. The most unusual part of the Grays' appearance is their very large, round, bald and disproportionate heads. However, in some varieties, a heads seemed to be elongated, or showed a strongly protruding nape. Some Grays had chins that were sloping, making their heads look like a light. However, X-ray examination clearly showed the upper and lower jaws, as well as the structures of the cranial bone. Forensic experts believed that the Gray's brains were divided into four separate lobes, and were disproportionately large compared to the human brain. Some have reported two regions of the brain separated by a transverse septum. It separated the anterior from the posterior, and outwardly they seemed to have no connection [3, 4, 5, 10].

According to descriptions of some witnesses, Grays had huge, scary penetrating eyes without lids. Bill Cooper noted that each eye was about 7.5 cm from corner to corner and 6 cm across in height, the abaxial ends of the eyes were oblique upwards by about 10 degrees and directed towards the nose. Leonard Stringfield also noted the same, the eyes were deep planted, had no eyelids, there was only something like a fold. Philip Corso wrote that the eye sockets of the Grays were excessively large. According to Judy Dorathy, the eyes seemed transparent, with some kind of lines in the middle and seemed to flash. Around them was a band of gray. The pupils and iris were not visible in the eyes. However there were accompanying descriptions. Pathologists who performed an autopsy on one of these creatures at the Walter Reed Hospital said that eyes of these anthrops were covered with a dark film, similar to contact lenses. They allegedly enhanced the light even in complete darkness. The images appeared greenish-orange, depending on how they moved. Some abductees also believed that eyes of these creatures were covered with some kind of contact lenses (the some other aliens of quite normal human growth also had such contact lenses) [3, 4, 5, 10, 14].

Apparently, contact lenses could have another device. Dr. Sarsbacher (see Part 1) after talking with colleagues from the Pentagon (employees of Vannevar Bush, Robert Oppenheimer and John von Neumann) got the impression eyelids of creatures close in the middle, and the width of the slit depends on the intensity of the light. Abductee Pat also recalled that when the creature's eyes blinked, its eyelids closed in the middle. William English described a vertical slits Gray eyes as a light yellow-green stripe. What constitutes a vertical formation in Gray's eyes is not entirely clear. As noted by Willam Cooper and Robert Emerger, if the creatures were not wearing contact lenses, their eyes showed a clear iris, apparently a light yellow, At the same time, there was also a «vertical pupil similar to that of a cat or a crocodile» [3, 5].

However, appearance of vertical pupils in humans is not surprising and it is hardly worth considering the Grays are «reptilians». Some persons with «cat pupil» have Schmid-Fraccaro syndrome. This pathology is associated with the presence of an additional fragment in the 22th chromosome. This syndrome is accompanied by other congenital features, in particular, dementia.

Some animals have a third eyelid, but it completely covers the eyeball and does not merge with other eyelids in the middle. Sometimes human rudimentary third eyelid elso develops to an overgrown atavism. This third eyelid, as a bird eyelid, is able to completely cover the eyeball. The figure shows such an eye in an open, half-closed and completely closed state with the third eyelid. Strangely, when closed, visible through the translucent membrane of the third eyelid, the initially round pupil appears vertical (for comparison, the lapwing eye is shown below, the pupil stay round!). The origin of this video and the object is unknown.

Famus «reptilians» Nicki Minaj and Zhanna Inanna

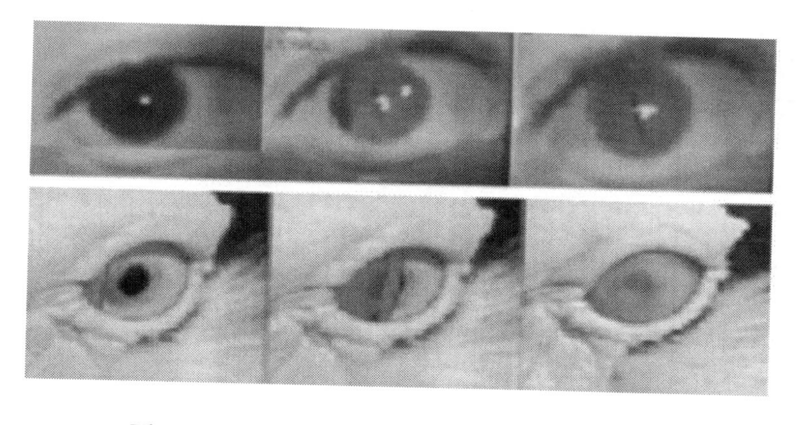

The transparent third eyelid of human (at top)
and the nictitating memdrane in lapwing eye (at bottom)

However, what constitutes a vertical formation in the eyes of the Grays is not entirely clear. If the creatures were not wearing contact lenses, their eyes showed a clear iris, apparently a light yellow, as noted by Bill Cooper and Robert Emerger. At the same time, there was also a «vertical pupil similar to that of a cat or a crocodile». However, even if the pupil was vertical, it is hardly worth considering the Grays «reptilians». After all, between people also meet some persons with «cat pupil», this is Schmid-Fraccaro syndrome. This pathology is associated with presence of an additional fragment of the 22nd chromosome. This syndrome is accompanied by other congenital features, in particular, dementia. The simultaneous appearance of such pathologies as the Schmid-Fraccaro syndrome and overgrown third century atavism is unlikely in natural conditions. However, the Grays may develop any multiple pathologies and deformities. And below we discuss why.

The Grays's skin was gray-brown or greenish, from which, in fact, this variety got its name («Grays»). They had neither hair nor hair follicles (like the aforementioned dogs fallowed to Chupacabrus on hunt). However, this information was not clear. Bill Cooper reported that the skin was described as thick and «reptilian». On the other hand, Leonard Stringfield noted that their skin seemed mobile, elastic. On the exposed parts of the body, she looked pale and transparent. Some clarity was introduced by Corso: Grey's skin was covered with a layer of waterproof fabric, which was very close to the skin [5, 10, 11].

The Gray's mouth was very small, like an incision resembling a fold. According to Leonard Stringfield, Grays had no teeth, ther tongue was atrophied. However, Michigan retiree William Curtis, who was abducted in 1974, noticed that the tongue was still there. After Curtis stuck out his tongue at a Gray, he opened his own mouth and showed his little tongue like in response (it is difficult to say whether they can fully talk). His toothless jaws were greatly reduced [10, 11, 15].

The Grey's nose was small, barely protruding from the skull. There were two nostrils on the flat face where it was supposed to be. However, in the drawings, the nose of some Grays, although small, is present. The Grays did not have a prominent chin. (However, this may be nothing more than an individual or a small group of individuals feature). There were no auricles, and the auditory openings were indentations passing into the canals of the middle and inner ear, the same as it is usually in humans [5, 10].

According to Corso, the Gray's hands were four-fingered, they did not have a thumb spaced apart [10]. However, Corso also mentioned six-fingered Grays, which gave reason to his detractors to accuse him of falsification. Abductee Angie, Anita's granddaughter, and Carla Turner herself also believed the creatures were four-fingered [3]. Four-toed aliens were also remembered by a girl Vera Potapenko from Orsk (1983) [16]. But Boyd Bushman claimed that the hands of Grays are five-fingered with fingers more elongated than those of humans and even showed a photo of such a hand [17]. In a sensational film released in 1995 by London entrepreneur Ray Santilli and depicting an «alien autopsy», a six-fingered creature was generally demonstrated. However, there is nothing strange about the different number of Gray's fingers. Syndactyly (reduction in the number of fingers) and Polydactyly (the presence of extra fingers) and other malformations of limbs are also found in ordinary people. Grays do not represent a specific biologic species. These are products of artificial cultivation, which can manifest any deformities.

The same contradictions exist in the description of the Grey's foot. Bill Cooper and Carla Turner noted that the fingers ended in claws and that the same four clawed fingers were on the feet. Interestingly, there were membranes between the fingers and toes [4, 11]. According to Stringfield, foot toes were missing, and the skin covered the legs in such a way that they appeared to be wearing socks. However, X-ray examination showed normal bone structure of the lower extremities [5, 6]. The legs of the creatures, according to many witnesses, were thin and short, with small feet. Thus, according to these signs, the Grays do not have a clearly established uniformity, and the spread of signs is very large.

Some of Linda Howe's informants have claimed that the Grey's skeleton resembled that of a bird. Flying birds are known to have hollow bones. However, according to Corso, the bones were thin and «fibrous» but more resilient than human bones. According to doctors, they were more like cartilage than solid bone. Some varieties of Grays had a very large chest, with more ribs than humans. And although their shoulders seemed very narrow, and their knees and elbows were not visible, X-ray examination of the skeleton clearly showed that these creatures belonged to anthropomorphs, no matter how strange they looked [5, 10, 11].

Regarding a blood composition of Grays, information is also contradictory. Philippe Corso, noted that Gray's vascular system is not closed, blood is mixed with lymph (which is very strange for vertebrates). This liquid was pink in color. However, expert studies were carried out on semi-decomposed material, which Corso himself recognized. It is unlikely that such features could be established under such conditions. It seemed that someone really wanted to draw parallels with insects, in which the circulatory system is not closed.

Leonard Stringfield claimed that the Grays had colorless blood [5, 6], that is, without erythrocytes and hemoglobin. This is strange, since pigments, hemoglobin, hemocyanin or hemoerythrin, etc. are present in blood of most chordates and invertebrates. In insect hemolymph oxygen carrier pigments are generally absent. However, the Grays belong to a chordates, it would seem that hemoglobin should be present in their blood. It turns out there are exceptions. For example, the «icefish», *Champsocephalus gunnari*, belonging to the *Channichthyidae* family, has neither red blood cells nor hemoglobin. Apparently, loss of hemoglobin is due to a genetic mutation, accompanied by loss of ability to synthesize hemoglobin and myoglobin. These fish survive in Antarctic in very low temperatures (1 – 2 ° C) and in oxygen-rich water. However, to increase an efficiency of oxygen transport, these animals have acquired a larger heart and blood vessels, their organs are permeated with a dense network of colorless vessels. Energy is saved due to the reduction of some organs (renal tubules, etc.) [18].

Perhaps Grays acquired similar adaptations. Philippe Corso wrote that although biochemistry of creatures did not contain any new basic elements, creatures had a very slow metabolism. In addition, autopsy experts noted that *their heart and lungs were also relatively larger than those of humans* [10]. We now know that the atmosphere of deep cavities inhabited by Grey's insectoid hosts is low in oxygen (12%). Surely, Grays are also adapted to such conditions, so an excess of oxygen on the surface can be toxic for them. Perhaps that is why they are forced to use masks to facilitate breathing [3, 4].

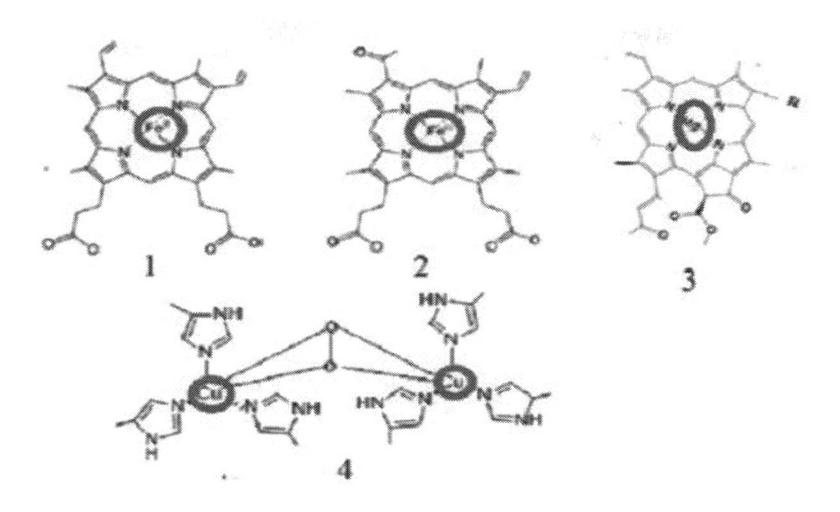

Prosthetic groups of Fe-containing Hemoglobin (1),
Fe-containing Chlorcruorin (2), Mg-containing Chlorophylls (3)
and Cu-containing Hemocyanin (4)

It has been noted more than once that the skin of the Grays has a greenish coloration. Williaml Cooper, who was able to look into the materials of the report number 13 of the project «Grudge», claimed that a substance «similar to chlorophyll» was found in blood of Grays [5, 11], on this basis it was even suggested that Grays eat autotrophically. The structure of chlorophylls and hemoglobins, in fact, is similar (magnesium is part of chlorophyll, not iron). However, this is complete nonsense, because even if photosynthetic cells completely cover the body, its surface area is completely insufficient to provide the body with energy. Grey's blood color could be due to hemoglobin analogues such as chlorocruorin or erythrocruarin. However, they are found only in invertebrates.

It is more likely that Gray's blood could still be stained with hemoglobin, or rather, its decay products, for example, biliverdin. In vertebrates, it is usually eliminated through a bile ducts and intestines. However, in some animals, for example, in five species of skinks of the genus Prasinohaema from New Zealand, as well as in one of species of frogs of Indochina, Chirixalus samkosensis, biliverdin is not excreted in the bile, but enters back into blood. Due to its high blood content, it is colored bright green. Perhaps the same situation is to Grays? After all, according to the description, their intestines are reduced (see below).

On the other hand, it is known that under certain conditions, hemoglobin can easily react with sulfur compounds, and the resulting compound, sulfohemoglobin, gives the blood a dark green tint. Sulfohemoglobin can be formed in a solution of hemoglobin when it is simultaneously exposed to hydrogen sulfide and oxygen (or hydrogen peroxide) or when hemoglobin rots. People whose body is oversaturated with sulfur do not experience much discomfort. Ralph of Coggeshall and William of Newburgh, reported on the sudden and unexplained arrival in the village Woolpit of Suffolk of two green children during one summer in the 12th century. Local peasants found them when they crawled out from some deep hole and got lost. The children spoke an unknown language and refused regular food. But the most surprising was their green skin tone. They were probably people from some underground isolate who were subjected to chronic sulfur poisoning. Gradually, the green tint of their skin disappeared.

According to Stringfield and Bill Cooper, the Gray's genitals were reduced, in some cases there were poorly developed female-type genitals, including two atrophied nipples on the chest [5, 6, 10]. These creatures are unlikely to have offspring. With such a small body, their huge head of the fetus generally makes natural childbirth impossible. So, *to maintain the population of Grays, artificial rearing methods must be used, including IVF or cloning, surrogate motherhood,* followed by growing the fetus in vessels with culture fluid [3]. (Here we come to understand the meaning of abductions and reproductive experiences of Aliens!).

But the most surprising, perhaps, was the absence of digestive organs in the Grays [6, 10]. Their oral cavity was described as a blind chamber about 5 cm deep, separated by a membrane from the reduced digestive tract. According to many testimonies, grays have underdeveloped toothless jaws [5, 15]. It was absolutely incomprehensible how such creatures could eat. According to some claims at a military base captured Gray allegedly has been feeding by ice cream and he absorbed this food through the «membrane». However, it is hardly possible. In the absence of a digestive system, even ice cream cannot be digested. The assumption that Grais can absorb liquid products and metabolites through the skin looks completely implausible. A creature without a digestive system could only be fed by intravenous injections of soluble glucose, amino acids, small peptides and other low molecular weight products.

Complete or partial reduction of some organs testifies about deep-seated symbiosis or parasitisme, when unclaimed functions of an adult organism are simplified, while maintaining some prenatal ones. If Grey's metamorphosis took place in this direction, then after the «birth» from the vessel with the culture liquid, nutrition could continue through the umbilical cord. If so, then the digestive organs are really not needed. However there was not a navels in lower abdomen in Grai's photographs seen by Bill Cooper [5, 11], although this part of the body could be hidden by a tight-fitting suit.

It is possible elso to feeding a nutrient solution into large veins through a catheter. It is not so difficult for Aliens, who master this technique (Abductants have repeatedly described thin hoses with needles. Various fluids were taken from them in this way: blood, lymph, etc.). Some Grays had bracelets around their wrists, probably with built-in catheters. Captured living Grays terribly screamed when trying to remove them. The stranger, discovered near Gdansk in February 1959, died shortly after this vital bracelet was removed from him [15, 19]. A blood-feeding and blood-purifying devices must be connected by this bracelets.

Strange Humanoids by Jader Pereira

Be that as it may, but Grai's anatomy and physiology have not any of extraterrestrial and any of insectoid (except conclusion of a «open circulatory system», that must be done erroneously). Complete or partial reduction of some organs is characteristic of deep-seated *symbiosis or parasitisme*. Based on external features, Grays do not seem at all to be any specific population species or even a supported breed (within which crossing is also maintain characteristics corresponding to this breed). Rather, the Grays are a collection of individuals within which there is no exchange of genetic material, and where signs are distributed very chaotically. At the same time, their far-reaching modification is evident. Probably Grays are Anthrops with the deepest modification.

In addition to small Grays, apparently in Alien society there are a bigger Grays. Abductants also call them «Grays», although they are clearly different from small ones. Polly, an abductee from the Turner group, recalled: «I ... sat in some ship behind the pilot... and looked at his bulbous head back. He turned to me, and it turned out that he was wearing a mask... «. Gray was the same height as Polly and was able to normally converse [3].

Others varieties of Anthrops with a greater or lesser degree of genetics modification are present in Alien Environement. In November 1974, the french magazine «Phenomenes Spatiaux» published a big article of the South American ufologist Jader V. Pereira. Он проанализировал 334 свидетельства очевидцев о «внеземных» гуманоидах и составил их обобщенные характеристики [20].

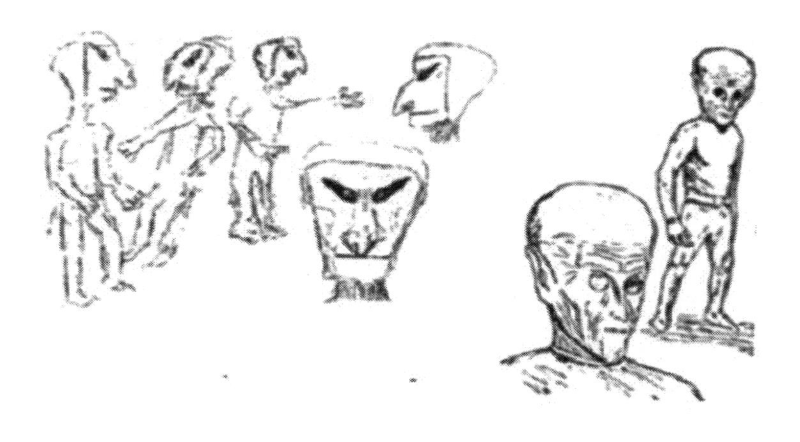

Nosed dwarfs. Drawings by R. Cooper [11] (left) and J. Pereira [17] (right)

In Alien Environement there are some Anthrops, which, unlike Grays, are quite capable of feeding on their own. They are engaged in theft of agricultural products. The list of food basket includes flour, oats, other products of cultivation and horticulture. In Wisconsin, UFO pilots treated farmer Joe Simonton to flatbread that looked like cardboard (1961). As it turned out after laboratory analysis, the cake consisted of hydrogenated fats, starch, buckwheat husks, soybeans and wheat bran, that is, products of completely earthly origin. In January 1979, strange creatures enjoyed homemade minced meat pie at Jean Hingley hause, a housewife from Rowley Regis in the West Midlands of the UK. And in August 1986, strange UFO passengers stole two chickens from a peasant's yard in Trieste, Italy [21 – 23].

William Cooper saw stranges dwarfs in the photographs of Project «Grudge». They were standing against a wall in an empty concrete room and looked extremely confused. They had huge hooked noses which large nostrils. Their almond-shaped eyes were set very obliquely. They were wider near the nose and tapered at the sides. The eyes had whites and dark irises with a vertical pupil. The pupil seemed to glow. Their mouth was wide, like a slit, the lips and teeth were invisible and mandible was wide. The head was larger than a human head used to be, the hands were small with thick fingers. In accompanying materials, it was noted that their skin color depended on nutrition. The legs were small, shod in shoe covers. The clothes were similar to overalls, they covered the whole body and the head from the sides. The ears were not visible. According to Robert Imininger, such dwarfs visited Holloman Air Force Base in 1964 [5]. It is difficult to say whether other dwarfs, also very similar to fairy gnomes, belong to the same variety. One abductee recalled being led to a little old man with pointy ears and a goatee. This creature had quite normal human eyes [14].

The Spanish ufologist J. Benitez told about the observation of two dwarfs who killed sheep in southwestern Bolivia (1968). Local peasant woman Valentina Flores took a good look at one of them. It was a small creature, about 1m 10 cm tall, with a completely human face. His skin was very white, his eyes and hair were blond, he had a thick red mustache and beard. When Flores hit him with a stick, he showed fear and pain [24].

Kidnappers of Xavier Clare's have took «selfie»

There are also completely ugly Anthrops. In 1985, the Spanish photographer Xavier Geres Clares was kidnapped in the Valgorginha area. The kidnappers pushed him into some kind of underground cave where it was difficult to breathe and allegedly even tried to take possession of his camera, taking «selfies» [25]. The pictures showed disgusting faces. Along with creatures that had a clearly human appearance and proportions, there are very large anthrop, two meters tall or more.

As many contactee witnesses say, the most ordinary guys and pilots have been repeatedly seen in UFOs and among Anthrops. In 95% of cases, Anthrops of normal human growth, very similar to us, were observed. Sometimes they were in spacesuits and masks, but they could do without them, they could be dressed in overalls, uniforms, etc. On a June night in 1988, a resident of the island of Puerto Rico, Carlos Manuel Mercado, was abducted by to an underground nest under Mount El-Cahul, next to Lake Laguna Cartagena. There he saw not only small bald Grays, but also a tall being of «human appearance». The little ones treated him like a boss [26].

Similar large Anthrop were seen by other abductees. They were met in Brazil, the Argentine, the USA, France, England and the New Zealand. This Anthrop people have hair that is short in men and down to the shoulders – in women, normal skin color, sometimes yellowish or dark. They can communicate with people, and even in contactee language (may be by «telepathy»).

Sometimes these «People» are behaving very strangely, inhumanly. They stood in front of Jane in white clothes, very tall and pale. They all stared at her and made a buzzing vibrating sound. One of them was blond, the others had darker hair [3]. Among the aliens there are also women who play the roles of doctors, nurses [3, 27 – 29].

ANTHROP REPRODUCTION and HUMAN ABDUCTION

Perhaps to better understand what pass in undergraund nests and in Alien society in general, should recall an ability of some social insects to enslave individuals of another species. So do, for example, ants of genus Formica, Raptiformica, Harpagoxenus and Temnothorax. They are slave owners, stealing eggs from ants of other species. Although a analogie method is not very scientific, and specific examples among termites are not yet known, however, general principles of building relationships among social insects are largely similar [30]. After studying what happens in Alien underground nests, such comparison suggests itself.

There is no doubt that Anthrops are a result of artificial reproduction of human spesies with some modification. In a conversation with abductee Amy, one Grays declared that he belongs to a special human race created by some unknown breeders, «creatures of a completely different nature» [3].

Probably, in ancient times, Aliens caught achildren of different human tribes, gradually domesticating them. The original forms of anthrops could also be quite archaic species of hominids. Some Antrops are completely covered with hair, like a bigfoot [20]. Their ancestral form could be the oldest giant astralopithecines. Along with these creatures there are Anthrops very like to modern people. It is possible that there are varieties of Anthrops that have undergone more modification than others. Small Grays seem to have the most profound genetic modification that was accompanied entire organ systems changing.

Over time, with the development of biotechnology, Aliens mastered new methods of genetic modification, using some types factors and vectors of transduction (for example, the «allocide-dependent transduction factor» is mentioned in REPORT-96). Therefore, appearance of strange freaks and discovery of no less strange mummies, that sometimes find diggers no longer seem surprising (Perhaps the stories about the Kyshtym dwarf, gnomes appearing in different parts of the world, in Western and Eastern Europe, Europe, South America are not fiction at all, but strange ugly mummies could be not fakes).

Breeders, using advanced biotechnology, are quite capable of creating a variety of deformities, but extraterrestrial biology has nothing to do with it.

Apparently, most pygmy anthrops are incapable of childbearing. Probably, IVF and surrogate gestation in the womb of ordinary women are used to reproduce the slave caste and abductions of women are carried out precisely for this purpose. Experts who unbiasedly studied this problem wrote that development of a fetus occurs in a womb of a surrogate mother for three months, after which the fetus is removed and grown in a vessel with cultural fluid. Captured women women rarely have memories of an unpleasant and dangerous adventure and don't supsone about an artificial insemination or about foreign germ implanted in their body. Some women are kidnapped and used for surrogacy multiple times. They, like sows, are kept in a cramped space in some kind of isolate for a long time. Aliens treat men else the same hidden, taking away their sperm [3, 4, 29, 31].

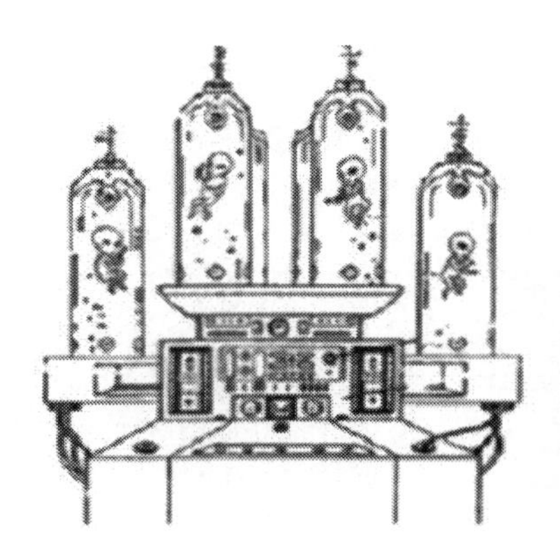

Anthrop embryos in vessels, drawing by Jane (left) [3].
Genetic disk (right)

Obviously a human genetic material undergoes some modification. It is possible that Aliens use some factors that cause directed mutations. Hereditary abductions of members of the same family clearly indicate some techniques adopted by breeders. It is quite obvious that «hybridization» of human with unrelated animal species is impossible. However, modification of human DNA with individual foreign genes from distant type of organisms by genetic engineering technologies could be carried out. The same technology is used, for example, for GMO-productions. Something very reminiscent was described by Fred (pseudonym). Then, in the reproductive procedure, not only ordinary men and women were used, but also an «hoofed animal» [4].

Evidently some Anthrops are reproduced by cloning. Stories have even become known that testify to the selection of organs containing stem cells. For example, one of them was told by Russian singer Katya Lel at a press conference on contact with alien civilizations, held on February 13, 2019 in Moscow. According to the artist, at the age of 16 in the Nalchik of Kabardino-Balkarian Republic she had direct contact with UFO. After this meeting, Katya lost some of her helth teeth, however she retained conscious memories of this removal process. The meaning of such operation becomes clear, given that a dental pulp contains stem cells suitable for growing clones.

The cloning technology have no many differs from growing babies from a fertilized egg. Advanced alien techniques certainly allow do it more successfully than it was done in sheep Dolly project of Campbell and Wilmuth (1996). The prenatal stages of development of the fetus and the clone should be similar, and babies born under artificial conditions will not differ much. However, growing an adult organism in a vessel raises many more questions. According to some of the abductees, «alien doctors» allegedly showed adult bodies in large containers filled with liquid. Goals of such experiments raises many questions (about a possibility of normal body development withaut learning and movement, about a rate of this process, not to mention the method of hypothetical relocation of the «soul») [3, 25, 28].

So this information from abductees does not inspire confidence. However, some Anthropes are definitely clones. People who had an experience of their abduction retained memories about the same Antrop types. Over the years, they did not change at all and were always young [27]. For example, a fair-haired young man with a characteristic appearance was encountered in a foreign environment in the 1950s [4, 28, 32] and in the 1980s [5] and even in the mid-1990s [3]. The second most common type is a tall brunette with very large eyes and characteristic bald patches («widower's line»). It was also present in the stories of abductees and contactees both in the 1950s and in the late 1990s [3]. Moreover, according to social surveys, from 2006 to 2011, many people retained conscious and unconscious memories of him (now they are trying to convince us that the site that published these studies used the project of marketer Andrea Natella) [33].

Surely the technology of breeding and raising Anthropes is well and widely established in a alien society. There must be corresponding textbooks and visual aids. An artifact, existing for a long time and so-called «Genetic Disc» found in Columbia is very interesting. This ancient 27 cm diameter disk was made from lidite by using an unknown technique of convex relief on. There are an images of a man, woman and human life cycle. Some of its details could be observed with a microscope only: the formation of a male spermatozoon and a female egg, fertilization and development of a human embryo at different stages. Such disk could be used as a teaching aid in the study of human ontogeny (It is interesting that similar stone with «teaching aids» were found in other places. They contain maps of an area, astronomical pictures, including images of constellations. In China they are called «Bi». The earliest of them traditionally attributed to the Neolithic).

Most likely, plastic figures depicting people of different ages, which Polly (a pseudonym) have been shown in an Alien Environment, are also studing aids. She visited Antrope kindergarten and saw babies and older children there. When she held a little blond and blue-eyed boy similar to her in her arms, it seemed to her that this was her child [3]. Some abductees were shown not only babies, but also children schools. Children were thin, fair-haired, dressed in the same light overalls, they were orphans. Perhaps Aliens do not accidentally invite weemen to visit heir biological child. Young children need such at least short-term physical contact [3, 31]. A similar procedure was described by a lady from the Ukraine. When she took hands a child that was suffering, he felt better [29, 31].

Anthropes take an active part in abductions and reproductive operations. As a rule, Anthropes with a normal human appearance are included in the team, because in their presence abductees feel calmer. However, most procedures are performed by Grays of various sizes and even by Upper Cast Aliens, who are usually leaders. That's how it was in David's (pseudonym) operation [4] and in other cases mentioned in the previous part of the book (see above). In 1980 in north Denver, a handsome young blond man lured a married couple onto a ship, where the Grays performed agonizing medical procedures on them. In 1985, in Minnesota, UFO pilots captured a woman working in a field. When her husband, a farmer, rushed to help, the blond-haired man tried to stop him, but the husband still ran to where his wife lay surrounded by small gray creatures [34]. Sometimes women – «medics» and tall, big-eyed doctors take part in the procedures [3, 4].

HUNTERS FROM UFO

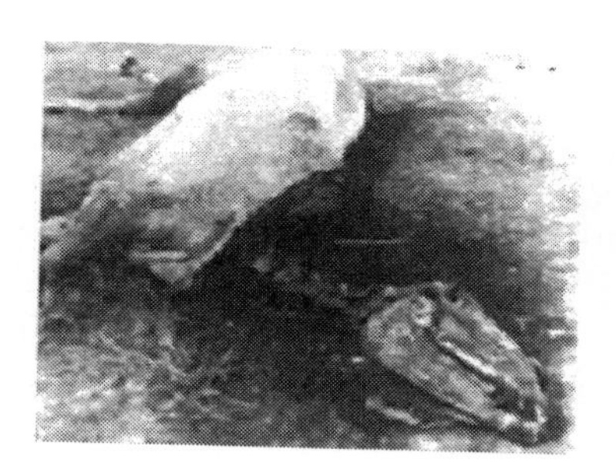

Mutilation victims: the horse Lady
and the bull Snippy (1967) [5]

One of the most important functions of slave symbiotes in Alien society is to hunt warm-blooded animals. It has the same purpose as Chupacabrus attack: it have to collect a blood product. Unlike chupacabrus attacks, this method of hunting uses technical means: flying objects, lifts, search tools.

This phenomenon has been known since ancient times. Back in the 1st century BC Chinese scientist Sima Qian wrote about a mysterious object called the «Heavenly Dog». The Chinese associated her visit with death, bleeding, theft of the internal organs of animals and people. The heavenly dog, «cheng-cheng», is also mentioned in the «History of the Southern Dynasties» for 514 AD.

Talks of strangely mutilated carcasses of cows and horses had been circulat among pastoralists of the United States longetime, but usually they were explaned as an attack by predators. When some people noticed that carcasses were mutilated with knives, these crimes were attributed to unknown sacred murderers. These cases began to be widely discussed in the 1960s. After half a century ago, a racehorse Lady, who died a terrible death, was found on one of the American farms, information about such incidents began to seep into the press. The first article about this was published in the Colorado newspaper «De Pueblo Chieftain Newspepie», October 7, 1967. These cases became widely known after the release of Linda Moulton Howe's film «Alien Harvest» (1981). It suggested a connection between strange phenomena: the appearance of UFOs, unmotivated killings of livestock [5].

Since the late 1960s, such phenomena have been constantly repeated in the pastoral farms of the United States and Argentina. Later unmotivated and bizarre killings of domestic and wild animals start to notice in many parts of the world: in the British Isles, in the USSR, in Japan, in Indonesia and in Thailand (2017).

The anomalous nature of livestock killings and their connection to UFOs have been reported by MUFON Idaho director Don Mason, journalist George Knapp from Salt Lake City and British television journalist Zichard Hall. Professional British veterinarians from the private organization APFU, led by David Cayton, conducted research on the remains of sheep and other domestic animals [35]. In the United States, scientists from LANL and NIDSci were studying these threatening phenomena. The FBI and ATF joined to this investigation at the federal level. Private non-scientific organization Bureau of Investigation, located in Kingston, dealt with this problem in Canada. Official experts were involved in investigations of livestock killings in the province La Pampa of Argentina [36].

Studies have shown that these murders have some specificity. Cattle were slaughtered at the age of 4-5 years. Exsanguination was carried out in all cases. As a rule, not only the blood, but also cerebrospinal and interstitial fluids were pumped out. Probably, some technical devices were used for this, since it is not easy to completely bleed an animal. For this purpose, the mutilators previously injected solutions of anticoagulants, opiates, and oxidol. In addition, other chemicals were introduced, because after death animal skin hardened, bones turned black and became porous and brittle, muscles became discolored, and a liver acquired a waxy consistency. Killers removed some organs and tissues: genitals, rectum, eyes, tongue, mucous membranes of lips, anus and rectum. Sometimes big internal organs were excised through deep round holes. According to microscopic examination the laser beam was probably a cutting tool.

Sometimes the tissues were cut by without cell destroing or nerve tissues and vessels were isolated very precisely. In those cases when the incision passed through the spine, it did not touch the vertebrae, but neatly passed between them. However, all cases of anomalous attacks on livestock have always been hushed up. The general public was unaware of the true extent of the phenomenon. According to Howe, in the world there are at least two thousand «anomalous» slaughters during the year [5].

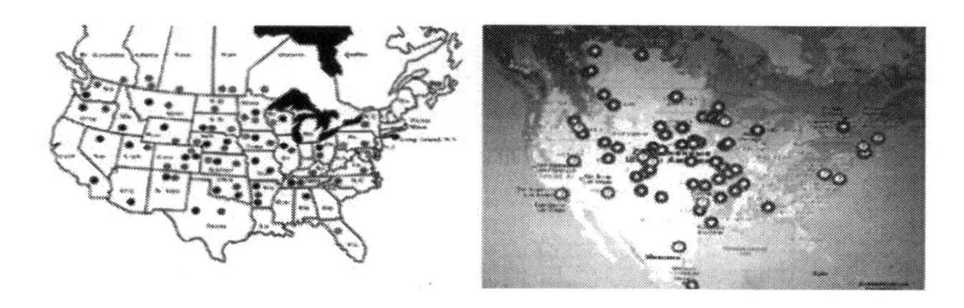

Mutilation sites in the USA by Linda M.
Howe (to 1989) [5] (left) and after 1989 (right).

Linda Howe's book contains a map of the United States showing the places where large mutilated animals have been found from 1967 to 1989 [5]. But it reflects a small part of the phenomenon, those cases only that got into open press. The next two decades, such points appeared much more. Numerous cases of mutilations have also occurred in South America, especially in the Argentine provinces La Pampa and Santa Fe, where many cows and other animals have died since 2002. Later these terrible cases occurred in the British Isles when sheeps were mutilated in the village Kainlohev of Scottish. In 2010, mass killings of sheep were repeated in Shrewsbury and Shropshire.

The scope of the phenomenon will seem even more menacing if we consider that not only livestock is subjected to mutilations, but also wild animals, which hardly anyone counts. Mutilated corpses of deer, bison, and grizzlies have been repeatedly found on the territory of the United States and Canada. In Alaska, near the Yukon River, tourists found many carcasses of wild animals and even a whale (1977). Mutilated seals were found on the islands of Scotland and on the coast of the Barents Sea (2002), dolphins – on beaches of France (1998) and the Galapagos Islands (2008). In 1997 and in 2010 corpses of dogs and cats cut in half were scattered in Virginia, New York, Massachusetts, Colorado, as well as in the province British Columbia of Canada [5, 37, 38].

Obviously, the slaughter of animals and the pumping of blood are carried out on an industrial scale. The blood is collected as food. First of all, it provides for physiological needs of Insectoid Aliens. By the way, ability to eat terrestrial products clearly indicates a terrestrial biological nature of Aliens.

Cases of anomalous animal deaths have been directly linked to UFO sightings. Numerous witnesses in the United States observed them hovering directly over pastures, where dead cattle were later found: over states Washington (1965), Colorado (1967, 1975, 1976, 1985, 2000), Nebraska (1974), Texas (1975), New Mexico (1976), Arkansas (1983, 1989), South Dakota (1984) and Missouri (1992). Locals saw a UFO over Long Island, New York, where the remains of numerous domestic and wild animals were subsequently discovered (1988). The same sightings of UFOs, associated in time with death of animals, were recorded in the neighboring Canadian provinces of Alberta and Ontario. There have been several such sightings in remote wastelands of Bodmin Moor in Cornwall (1996), pastures Rio Grande do Sul in Brazil (1972), and the Argentinean province of La Palma (2008), as well as isolated cases in the Japanese prefecture of Saga (1991), the Adelaide Hills in Australia (1984), in the village Amga of Yakut (1996), and in the Urals (1998) [5, 39, 40].

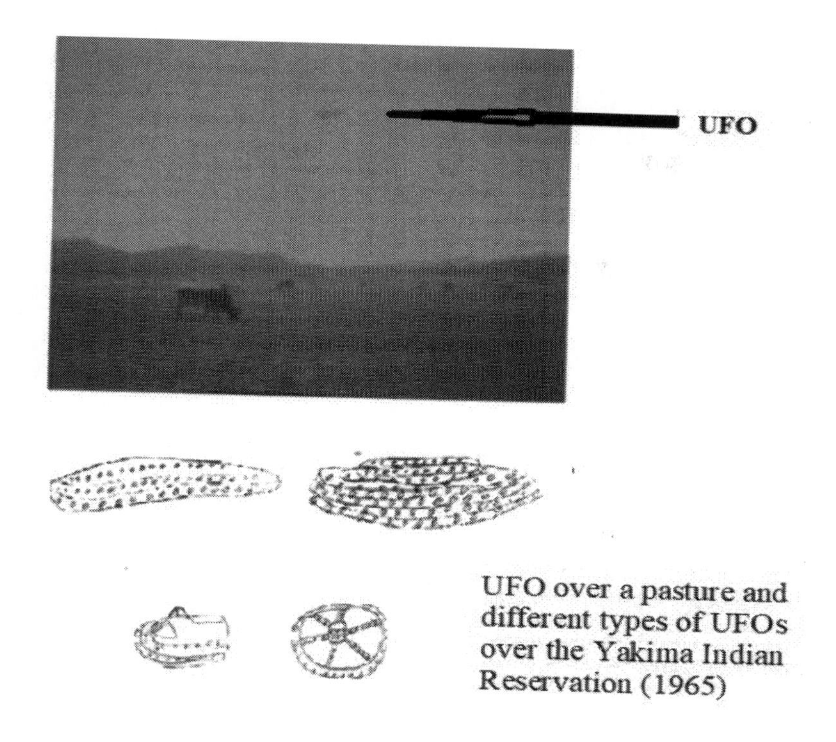

UFO over a pasture and different types of UFOs over the Yakima Indian Reservation (1965)

As a rule, a very large, well-lit, elongated flying object was the first to appear above the grazing areas. Small disc-shaped UFOs flew out of it like shuttles. There could have been five or six. The large object disappeared (or extinguished the lights) while the smaller ones sank low towards the pasture. Some time after the capture of the animal, disc-shaped UFOs quickly rose up and returned inside the large ship. This pattern was observed by residents of the Yakima Indian Reservation in Washington state (1965), residents of Colorado Springs of Colorado (1976-1977) and South Dakota (1984). Approximately the same observations were made in the west of Great Britain in the county of Shropshire, where, according to Phil Hoyle, smaller objects separated from the mysterious spheres, they descended to pastures, where mutilated mutton carcasses were later found (2010) [5, 35, 39].

Going down to the pasture, the discs hover low, leaving a round trail on the ground. These mysterious circles are nothing more than a trace of the influence of high-frequency UFO engines on the soil. Exposure to such radiation leads to heating of the soil, destruction of part of the microflora and burns of plant roots. Inside the circle, an increase in the level of radiation is recorded. More complex crop drawings have a completely different origin [40]. Often mutilated carcasses and even traces of supports were found near such circles [5].

UFO-nauts lift a victim aboard a flying object in various ways. In some cases, they use ropes with cat grappling hooks or throw a net over the animal. There are known cases mentioned on the pages of MUFON in 1999, as well as the Belarusian magazine «Incredible World» (early 90s), when animals were raised into an object with the help of some substance similar to fog. However, the most commonly used is an unknown force radiation, resembling a suspension of small shiny particles. With a similar force beam, not only a calf, but also contactee Judy Dorothy was lifted aboard a UFO (1973) [5] .

Humanoid creatures have been repeatedly seen near landed objects and places where dead animals were later found. Two killers were seen at close range about a hundred meters by a farmer who lived near Waco, Texas (1980), wich was looking for a lost cow ready to calve. He saw two Grays about 120 cm tall, they were dragging a calf. They were dressed in khaki overalls and with dark eyes. A couple from Missouri watched through binoculars two small creatures in silver suits dragging a paralyzed cow into a UFO (1983). In southwestern Bolivia, two dwarfs killed many sheep in a paddock of local resident Valentina Flores (2001) [5, 24]. She described one of them as a small, blue-eyed man with a red beard. He cried when she hit him with a stick.

Kidnapped Judy Dorathy (pseudonym) (1973) and Myrna Hanson (1980) described the killing of animals that took place right on board a UFO, in a cramped round room. Their memories were restored only after regressive hypnosis. Judy reported that two 90 cm tall dwarfs performed all operations aboard UFO. The victim was not killed immediately, but blood was pumped out of a still living animal and organs were cut off, ignoring his cries. They wielded long ribbon-like knives, cutting off eyes, tongue, udders, reproductive organs (in other cases, laser knives are used). After that, the mutilated body was thrown away. The dwarves distributed the cut tissues and organs into various containers.The mutilators collected body parts not only from cows, but also from other animals. One of the dwarfs quickly did something in an apparatus device, on which the lights were blinking. According to Judy, he tested products for contamination with poisons. (Most likely, we could talk about insecticides. This problem became relevant in the second half of the 20th century) [5]. The little dwarfs, about 90 cm tall. Their heads were large. Their eyes without eyelids, were penetrating and scary. They seemed transparent, with some kind of lines inside and with gray stripe around. Something changed all the time in their eyes, as if flickering and rotten. The skin was white and pasty. They had long crooked fingers with black claws. The shoulders were very small, as if without bones. They were reasonable and talked by meowing, similar to sounds of Chinese speech.

According to Judy, the mangled carcass was ejected from a UFO. It is because of falling from a height that many crippled animals have broken bones. Sometimes traces of a fall remained in the form of broken tree branches, and at a height where a large ungulate animal could not climb on by them-self. Some remains were found in «impossible places», on rooftops or pillars. Sometimes live cows were found on the roofs of houses or barns. It is possible that these were potential victims who were dropped on the way to a hovering flying object.

Victim of mutilators, dropped from a height

Although Myrna Hansen's convoluted regression protocol do not included a detailed description of mutilators, she did manage to recall some very interesting details. Being in some underground nest or base (presumably near Dulce, see part 2), Myrna accidentally got in a dimly lit room. There were metal tanks in which animal and may be human body parts floated, an unhealthy sweetish smell sented like pickled pork [5, 41].

In most cases, descriptions of mutilators matched a Grays, but not always. The mutilator from Bolivia appears to be more humanoid than Grey. It is possible that not only dwarfs, but also larger Anthrops are engaged in hunting from flying objects. In the summer of 1993, near the Mexican city of Los Andos, shepherds saw a UFO land in the middle of a herd of cows. Two-meter creatures came out of there and began to cripple cows right on ground. One shepherd went there on a motorcycle and was also killed [35, 42].

Apparently, hunting vampire gatherers is the original natural way of obtaining food in alien society. The hunting cattle using UFOs, carried out for the same purpose to provide nest inhabitants with a blood and other products. Compared to hunting Chupacabras, this method is much more efficient. There is reason to believe that hunting cattle using UFO is a specialization of Anthropes.

After all, Upper Aliens can't stay on the surface for long due to oxygen poisoning. It is unlikely that they are inclined to hard work and risk, exposing themselves to the danger of being shot down by human cannons. But Anthropes for Aliens are consumables, which is not a pity.

It is unlikely that foragers Chupacabrus are able of doing this job. Butchering of animal the bodies, especially organ removal with surgical precision, is very complex operation. Although Chupacabrus, compared to other animals, are distinguished by certain ingenuity, they are from worker or soldier castes of Aliens. They are capable of skillfully opening cages with rabbits, tearing the victim with their claws and sucking blood. It is also difficult to imagine that Chupacabrus could perform analyzes using instruments or control such a complex mechanism as a flying ship or even a shuttle. But dexterous humanoid slaves are quite suitable for this.

Apparently, hunting by vampire foragers is a original natural way of obtaining a food in alien society. It can be assumed hunting by UFO, much more efficase, appeared only after evolutionary development of intelligent anthropoids, Homo sapiens or their intelligent predecessors (that is, no more than 200 thousand years). These wild people had to be captured and domesticated. Perhaps only after this Alien and Anthrop symbiotic socium achieve technological progress.

HUMAN VICTIMS

On the part of Aliens, it is quite natural to treat people as slaves or experimental animals. They can be caught, propagated, gutted for alien needs. Many captured human victims do not came back, falloved the same way as captured animals. UFO-nauts act with extreme cruelty. We know this from surviving witnesses, for example after failed takeover attempts. Brazilians Ermelido and Antonio Amador de Lima were tried to capture with grappling cat hooks (1976), Gianunzio – with a power beam (1979) [43]. A similar case of attempt to capture a man occurred on the state farm «Voensovet» in the Rostov region of Russia (1989) [44].

Human mutilations also do not differ from what is done with animals. One of the earliest known cases of human mutation occurred on January 15, 1947. Then the mutilated body of 22-year-old Elizabeth Short was found in Leimert Park, not far from Los Angeles. Her body was bled and cut into two pieces at the waist (a similar murder occurred three decades later in Mexico – see below). The incision was made in such a strange way that the vertebrae were not damaged. Killers removed external and internal genital organs, nipples. The mouth was also mutilated. This case was investigated by FBI, but the killer was never found [45]. Journalists struggled with this riddle for a long time. No one guessed what was happening even when bodies of four bloodless victims were found in Yugoslavia in the early 1950s [46].

According to Philip Corso served in the intelligence service and had access to some classified documents, American authorities have been aware of murders, kidnappings, mutilations and blood pumping since 1951. Linda Howe elso was given important information by William English, one of the former employees of the US Air Force. He allegedly saw report number 13 of the US Air Force UFO Project «Grudge». It contained photographs not only of mutilated animals, but also of people. [47]. English met personally with the author of this report in 1977, when he served in the US Air Force in the UK. He told about a terrible incident in 1956 at the White Sands test site in the United States. One of the servicemen saw how a hovering UFO dragged Sergeant D. Lyett, and three days later in the desert they found his bloodless body with carved out eyeballs, genitals and rectum [5, 11].

One frightening episode that occurred in India in 1958 was reported by the English UFO researcher Jenny Randles. An Indian businessman, who wished to remain anonymous, said he saw a UFO land near two playing boys. Four creatures about a meter tall came out of it. Soon one of the children was found dead. He had several organs surgically removed. The second was in a catatonic trance and died five days later without regaining consciousness [48].

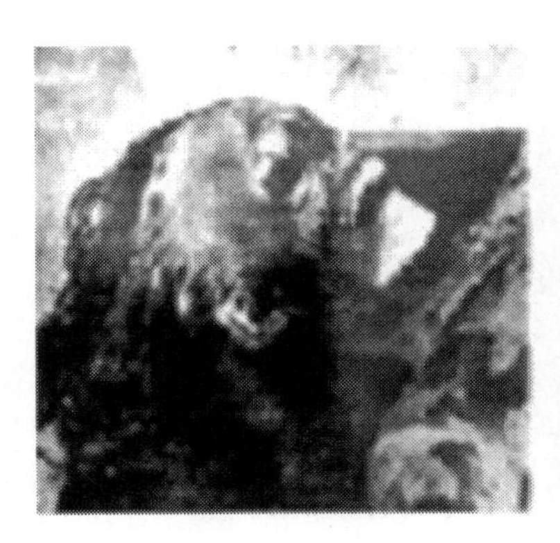

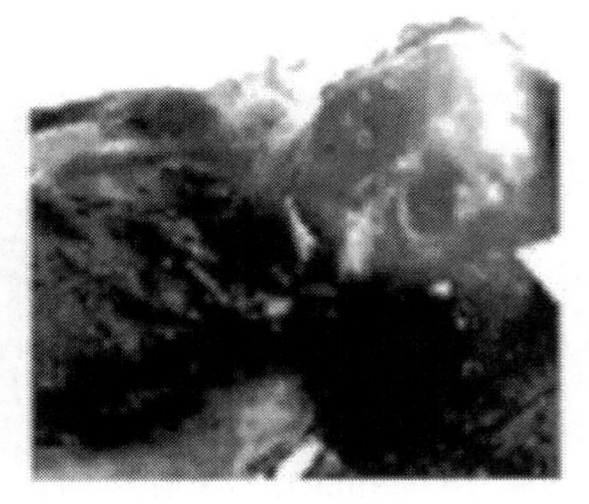

Remains of Lyudmila Dubinina and Semon Zolotarev (1959)

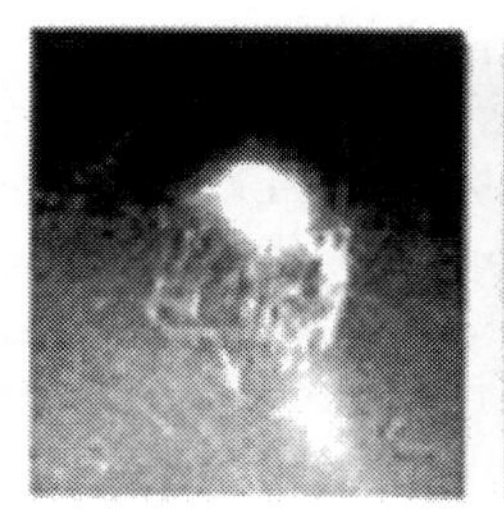
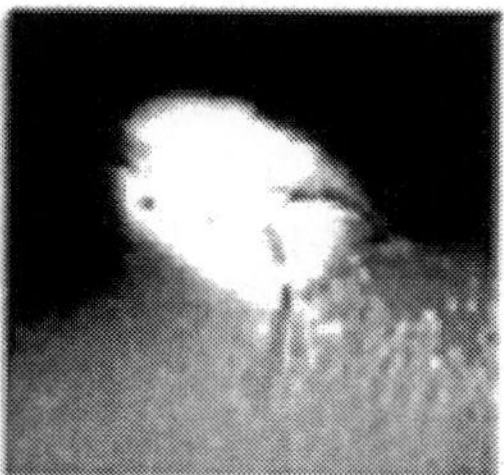

An flying object that was approaching to the Dyatlov's group tent. Stills from the camera of Semon Zolotarev

In early February 1959, in the Urals, on the way to the Otorten and Oiko-Chakur mountains, a group of tourists led by Igor Dyatlov died. There were seven boys and two girls. All died, their frozen corpses were found by a search party. Five bodies were found quite quickly, and four bodies, of which three were especially injured, were found only after 4 months. (Probably they were hidden away by a clearing team – see below). The version of mutilation is confirmed by very strange wounds that were found in these three bodies. Rustem Slobodin had a depressed skull fracture, which could only be the result of a fall from a great height. Probably the capture attempt failed. Semyon Zolotarev and Lyudmila Dubinina had broken bones and elso. Their eyeballs absented, Dubinina had her tongue cut out and her hyoid bone broken. According pathologists Dubinina's broken ribs should have caused extensive internal bleeding, incompatible with life. However, the examination indicated only minor hemorrhage, the body must had been bled dry. The conclusion of the histologist wich could confirm it in general suspiciously disappeared from the file. The clothes of these three victims showed distinct traces of radiation, which could indicate a possible contact with a UFO. Not far from the scene, searchers noticed a trace of a melted circle in a snow and burns on trees. The fact that a UFO could land there was also evidenced by the broken tops of young fir trees.

An approaching luminous object was captured on the last film frames of cameras of Yuri Krivonischenko and Semyon Zolotarev. It looks like a glowing ball. It should be mentioned that local Mansi residents repetedly observed strange fireballs over the place of death of tourists. They not only described this phenomenon, but also sketched it (subsequently, the drawings disappeared from the file). Many residents of the Northern Urals in February – March 1959 saw them many times. During the search operation, on February 17 and March 31, fireballs in the area were observed by rescuers [49].

Obviously, the order to classify this investigation came from Moscow. They well understood a danger of thiese incidents. Back in 1952, the American authorities warned leaders of major countries, including the USSR, about UFO killers. According to David Cayton in 1959, NATO leaders received 4 reports from the Soviet authorities [42]. Apparently, in that year, Soviet servicemen tried to clean up the place of mutilation by they-self, but the operation was not carried out enough «qualified», and many questions remained for the investigation (Later, this prerogative will be given to NATO troops – see below). Attempts to hide the circumstances of the death of Dyatlov group continue to this day. Not so long ago a RF prosecutor's office resumed the 60-year-old case. Everything is done in order to falsify documents and circumstances. Be that as it may, among ufologists (outside of Russia), this case is considered a classic mutilation.

In April 1972, an American special group operating in Cambodia came across mutilators while they were sorting and putting human body parts into special containers. A fight ensued, as a result of which several commandos died, having received burns. The Mutilators retreated to the UFO, also suffering casualties. After returning to the base, the members of the group were detained, they were interrogated, psychologists worked with them. The case was classified. A similar incident happened in the early 70s in the jungles of Vietnam. A B-52 bomber was found there, its crew died. People were tied to places and mutilated [6, 50]. It is not clear whether we can talk about the same case, since their circumstances are similar. Last event was told by William

English, who allegedly served as part of MACV/SOG search and rescue team in Vietnam.

In November 1977, bloodless corpses of three fishermen were found on one of islands of Lake Winnipeg in Canadian province of Manitoba. Local residents saw UFO over the lake last night. This case was classified.

In the same year, several disturbing cases took place in Mexico. In San Luis Potosi, there have been several cases of exsanguination of children associated with UFO sightings [51]. Early 1978 in the state of Tabasco, a heavy object suddenly fell on a car carrying «Petroleum Mexicana» employees. It was the torso of a man who fell from somewhere above, breaking the windshield. The human body was cut in half. The lower body was found far away in a field. As the examination established, the incision passed through the spine, but the vertebraes were not damaged. The remains were exsanguinated. As it turned out, it belonged to a local resident. The circumstances of his death could not be established. Salvador Freixedo, who investigated the case, concluded that both parts of the body were thrown from above. Interestingly, in the case of the death of Elizabeth Short in Los Angeles in 1947, the murder was carried out in the same way (see above). And not only then: In 1997, 2004 and 2010, in the states of California (in Long Beach), Virginia, New York, Massachusetts, Colorado and in the Canadian province of British Columbia, bodies of cats and dogs were found cut in half elso. The cuts were just as even and precise. These cases were also associated with UFO sightings. And in 1949, when a soldier dismembered in the same way was found in the Kobyakovsky caves of the Rostov region of Russia, no UFOs were reported [5, 37, 38, 52 – 54]. However, the «handwriting of the killer» was still the same.

In 1979, hunters found a body of a naked man in woods near the Bliss of Idaho. The man was mutilated and bled, his genitals and lips were cut out. There were no marks around the body. The man's bare feet left no footprints, no dirt on them, although the ground around was wet and sticky. The clothes were found ten kilometers from the corpse. There have been many reports of UFOs and cattle mutilations in the area [55].

In September 1988, in the Brazilian state of San Paulo, a fisherman body with signs of mutilation was found near the reservoir dam in Guarapiranga. It was an 53 years old man, who lived with his family near the reservoir. An autopsy showed cerebral edema and cardiac arrest due to pain shock. This gave grounds to assert that the injuries were inflicted during the life of the deceased. There were holes measuring 1-1.5 inches in diameter on the chest, abdomen, legs, arms and forearms. Internal organs were removed. The scrotum, testicles, and anus were removed through elliptical incisions. The tissues around the jaws were cut along with the lips. The ears, tongue and neck tissue were also removed. There was any rigor mortis. The corpse was completely bled, there were no signs of decomposition [56]. The Brazilian government, fearful of panic, tried to discredit the case by bringing in prominent ufologists. In order to reduce public excitement, an additional investigation was launched. Notorious spouses Kovo and Tanya Kunya took part in it. However, this raised even more questions [57]. Ufologist I. Z. Garcia refused to accept the results of the second study, since its custom nature was obvious.

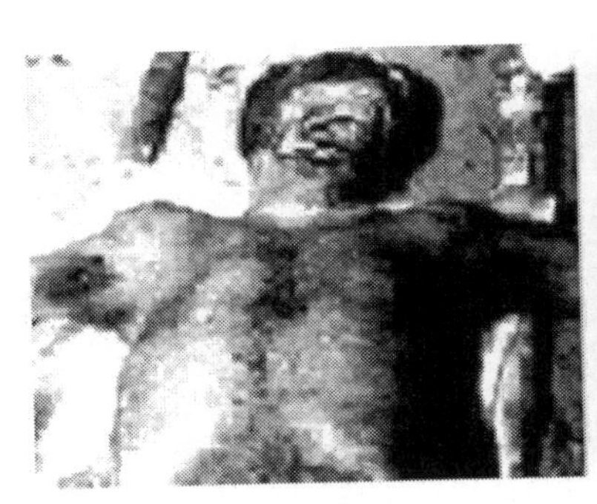

«Brazilian corpses»: from Guarapiranga (1988) (left) and unknown (right)

However, this makes no sense, since in these places in Brazil these cases have been repeated many times. A photograph of another Brazilian victim show missing eyes and disfigured lips.

At the same time, the burns on the forearms are located in the same way as in the man who died at the dam in Guarapiranga.

In 1989, in Argentina, the head of a police patrol approached a landing UFO and was dragged into the facility in front of his subordinates. Three days later, his body was discovered, from which all the endocrine glands were excised. A similar incident happened in the Mexican state of Chiapas. UFO landed among a herd of cows and two-meter creatures emerged from it. They began to trap and mutilate the cows. The shepherds watched from afar, one of whom went there on a motorcycle to prevent killers. When UFO flew away twelve crippled cows and a shepherd were found at the landing site. His genitals, pancreas and thyroid glands have been removed, the body was exsanguinated (1993) [42, 58]. On August 13, 2002, in state of La Pampa, police found the desiccated corpse of a man hanging from a large tree. The body was mutilated: on the left side, an eye, an ear and a piece of skin on the jaw and genitals were missing. The pathologist indicated that the incisions were identical to those usuly found in mutilation cases common in the Western LaPampa area. In the same year, 27 cases of cattle mutilation were reported in the vicinity of Córdoba [59].

Reconstruction of mopping up scene after mutilation in South Wales (1997). Frame from the movie «Silent Killers»

A terrible tragedy occurred in janvary 1997 in Britain. Young people, a 21-yers boy and a 16-year-old girl were killed on the road north of Cardiff in South Wales. They were horribly mutilated, bloodless, without genitals. The guy was skinned, his genitals were his genitals were removed with a smooth circular incision [42]. (The same kind of cuts were mentioned more than once: they were also in a corpse in the Brazilian Guarapiranga – see above, as well as in some mutated animals, bulls and seals)

39-year-old Todd Sis died in the Northumberland of Pennsylvania not far from his house. According to decision of Gary Steffen's Northumberland County Police, his body should not be shown to his relatives. Sergeant Kottner from Point Township police station let slip that there were some «strange things» associated with UFO (2000). The FBI joined the investigation. This case was reported by Bach Witkowski, director of the Pennsylvania UFO Research Center (UFORCOP) [60]. B 2003, a policeman was mutilated in Oregon. Eyewitnesses saw a small fireball rise above the scene. NATO's 58th Special Forces [42] took part in clearing the area and concealing evidence (see below).

Probably few people thought about the possible connection between dismemberment murder of 30-year-old Alexander Yushko (known as rapper Andy Cartwright) with mutilations. On July 29, 2020, St. Petersburg police found his body. According to the examination, the body was completely bled, divided into fifteen parts. At the same time, some internal organs were not found: stomach, pancreas, intestines, gallbladder [61]. The suspect was his wife Marina Kohal. The woman pleaded not guilty.

A special group of incidents are crimes attributed to Satanists. No one takes into account their possible connection with mutilators and UFOs. On April 16, 2005, five boys aged 10-12 disappeared in Krasnoyarsk. Three weeks later, the bodies of four of them were found in a sewerage. The fifth boy was never found. Three were completely bled, the fourth – partially [62].

In the same year, three families were mutilated in the Beni Mazar region of Egypt [50]. A similar incident occurred in September 2011 in Sevastopol, where two girls were killed and bled to death [63].

Mutilators showed interest not only in living people, but also in fresh corpses. It is unlikely that blood can be drained from a dead body, but there are known cases of removal of body parts from the dead. Such event took place in the province Alecante of Valencia. A deceased woman who was being prepared for burial, at night someone cut out eyes, tongue and amputated part of her arm. Immediately before bodies of ats and dogs were found in the basement of this mansion (1960s). In 1988, «fresh» corpses arrived at in Westchester morgue (st. New York) were mutilated by someone at night when the number of employed peaple was minimal. Parts of faces, eyes, thyroid and gonads of the corpses were removed. The results of investigation were not made public. Similar incidents were repeated the following year at two Connecticut morgues [5, 64].

Information about the attack of mutilators on people is one of the most classified in the world. *Some «supreme power» is trying by all means to prevent information dissemination about abnormal violence against people.* Many well-known cases have been discredited. Law enforcement agencies, as well as unscrupulous ufologists around the world, are involved in falsifications that discredit cases of mutilation of people. Although many of the above cases are well known to anomaly researchers they are considered «unconfirmed». Many well-known ufologists knew about the problem of human mutation, but were silent. Moreover, since some time it has become «bad form» to talk about mutilations of people among ufologists. Allen Hynek did not want to mention mutilation cases in his book «The Night Siege» (1987), written with co-authored by Philip Imbronier and Robert Pratt. According to Imbrogno, Hynek knew about the involvement of UFO pilots in abductions and murders of people in the Hudson Valley. Linda Howe also did not want to discuss these cases, at least publicly, although since 1975 she knew that they were real [5]. However, other, independent researchers have not been afraid to explore this issue. Don Ecker reported about such incidents have occurred in Europe, Mexico, Vietnam, Central and South America [55]. Ecker spoke about the reality of cases of human mutilation at the lecture «The Possibility of Human Mutilation». He mentioned human body parts found in Oregon (in the 1940s) and Texas (in the mid-

1960s). He noted that the search for information on the topic of human injury is very difficult: witnesses are intimidated, and most researchers do not want to get involved in a scandalous topic [65 – 67].

Russian ufologist Vladimir Georgievich Azhazha (1927-2018) spoke about cases of human mutilation that took place in Chile and Russia in a report at MUFON symposium in the Albuquerque of New Mexico (July 11-12, 1992). In his opinion, on the territory of the former USSR, missing people number as a result of UFO hunting reach 5 thousand persons for year. Peaple that are living or working in remote areas: farmers, geologists, sailors, tourists are in risk. Some of the missing were later found dead and mutilated.

At the 1997 UFO congress in Las Vegas, Professor Ademar Jose Gevard, founder and director of Brazilian Center for UFO Research, stated that in five years (1988-1993) 12 human corpses with organs removed were found in Brazil. He showed police photographs of two such corpses. The mutilations resembled the mutilations of cattle: eyes, lips, soft parts of the face, ears, and so on were cut out.

Alfred Beilhartz (1). Jared Nigret (2). Denis Martin (3)

Missing Children in US National Parks

In March 2007, American researcher Scott Corrales published sensational materials he had collected over several decades about human mutilations. It's no surprise that this month's blog posts have been removed from the Internet. Former San Jose police officer David Polides has collected data on numerous cases of disappearances of people in the national parks of America, which can not always be explained by accidents, animal attacks or crime [68]. Not trusting official statistics, he determined that every year thousands of citizens disappear without a trace in the United States and no one finds the remains of these people. Each disappearance story is strange enough, but similarities are found in hundreds of cases. This often happens in national parks. The vast majority of the disappeared are children. Adults also suddenly disappeared in there. Strangely, but National Park Service, law enforcement, and the media do not keep any statistics on such cases. Moreover, they openly lie or hush up the circumstances of disappearance of people. Don Acker also wrote about the mass disappearance of children in North America [65 – 67].

In 2014, noted researcher and television journalist Richard Hall released a sequel to «Silent Killers» [42], the first episodes of which focused on animal mutilation, which had been released five years earlier. The film features an interview with David Cayton, director of British organization APFU (Animal Pathology Field Unit) and veterinary experts who took part in the work of this organization.A new series is focusing on the problem of human mutilation. Caytoh listed a number of human incidents that took place in the 20th century. According to the authors of video, cases of anomalous killings of people occur around the world. However, they are the most secret topic in the politics.

The film also contains very curious and scary information about existence of special military units whose task is to monitor such cases. This is a story of one of retired soldiers living in the Breacon-Beacons of Wales. From 1983 to 1999 he served with the 58th Rapid Response Team stationed in Harriford. It is the so-called «Black Ops group» («Black Operations Group»). It consisted of 11 people, specially selected volunteers, who were on duty 24 hours a day. This service was created in NATO and was engaged in monitoring areas where UFO activity is observed. Near the Breacon-Beacons, an underground base was equipped with a special radar antenna 50-60 meters high, pulled out of the ground at night to track UFOs: the radar antenna rotated only at night. The same purpose was served by an underground structure near Sennybridge, there were 10 levels underground. A former member of the «Black Ops group» claimed that similar UFO tracking bases were being built in different countries of the world. The early warning system worked very reliably, accurately and quickly determining the coordinates of the place where the group should arrive. Theyrs group was always the first to arrive on the scene and the last to leave.

Each year, the group flew on a mission at least eight times. Having received the order, they quickly boarded a military transport aircraft, in the cargo compartment of which there was a helicopter. Upon arrival in an area, the military flew directly to the scene in the helicopter. Their duties included the protection of places of human mutilation. They quickly cordoned off the scene and put under guard everything that was there: human bodies, fixtures, etc. *However, they did not intervene in the «process» if mutilators were caught at the crime scene (!).*

In total, during the service of this military man, three to four dozen crippled bodies were found. This happened in the UK, Ireland, Scotland, Spain, Germany, Alaska, Australia, Yugoslavia and Russia. The most revealing case occurred in Australia, when 24 human corpses were found at the scene. Injuries in humans were the same as in cattle. The internal organs and brain were removed. Sometimes there were children among corpses.

These operations have always been led by the US military. A former employee of the Black Operations Group, among other things, noted that «the Russians left the sweep to the NATO teams that had the best equipment» (Apparently, this «reassignment» was apparently made after the unsuccessful sweep of the site of the murder of the Dyatlov group in 1959 G.). All materials from the places of human mutilations were usually taken by American scientists and forensic scientists. In some cases, the soldiers of group No 58 had to wait up to five days until they arrived at a remote scene. This group of 58 people was always the first to arrive on the scene and the last to leave [42].

Many questions arise in connection with this account of a former British soldier. And the most important of them: who is defending the armed forces of developed countries, we, the people, or the Aliens and their collaborators? Who really benefits from keeping us in the dark about what is happening here on our native Earth?

NOTES to Part 4

1.»Disclosure leaked ufo alien case video confidential documents old footage», 2011

2. «Briefing Document: Operation MAJESTIC 12» 18 Nov 1952

3. Turner K «Taken: Inside the Alien-Human Agenda», 1994

4. Turner K «Into the fringe», 1992

5. Howe L «An Alien Harvest. Further Evidence Linking Animal Mutilations and Human Abductions to Alien Life Forms», 1989

6. Stringfield LH «Status Report I – Retrievals of the Third Kind: A case study of alleged UFOs and occupants in military custody «, 1978

Stringfield LH. «Status report II – The UFO Crash/Retrieval Syndrome: New Sources, New Data», 1980

Stringfield LH. «Status Report III UFO Crash/Retrievals: Amassing the Evidence», 1982

7. Merezhko E «ARGUMENTS OF THE WEEK» 12/23/2017 (in Russian)

8. Reimink Troy. «In 'Bob Lazar: Area 51' documentary, director investigates UFO whistle-blower's story». Freep.com. Detroit Free Press. Retrieved July 31, 2019

9. Pravdivtsev VL «In the Land of Dreams». TOP SECRET, 05/01/1999 (in Russian)

10. Corso Ph. J, Birnes W. «The Day After Roswell», 1998

11. Cooper W. The secret government. The Origin, Identity, and Purpose of MJ-12. Quest Publications International Ltd, 1989/1990

12. «Briefing Document: Operation MAJESTIC 12» 18 Nov 1952

13. «Disclosure leaked ufo alien case video confidential documents old footage», 2011

14. UFO Magazine, 1987, 2, 4

15. UFO Magazine, 1987, 4, 3

16. Obiedkov O «Ufoids over Orsk», 1990 (in Russian)

17. «Cosmic Disclosure: Boyd Bushman's Death Revelation» divinecosmos.e-puzzle.ru/page.php?al=330

18. Kock KH // POLAR BIOLOGY. 2005,27, 862-895

19. Blania 3 «Sky full of UV» // PRZEGLAD TECHNICZNY, 1977, 28-44 (Poland)

20. Pereira JU «Les «Extra-terrestres»// GEPA, 1969, 2, P. 3-72

21. Valle J «Passport to Magonia : on UFOs, folklore, and parallel world», 1993

22. Randles J, «UFO study. A Handbook for Enthusiasts», 2007

23. ZA RUBEGOM, 1989, 1 (in Russian)

24. Galindez O. «Violent humanoid encountered in Bolivia»// FSR, July-August 1970

25. Ribera A «The Jinn and the Dolmen: The Most Amazing Case of Abduction Yet» // FSR – 1986, 31/4

26. Alien bases in Puerto Rico – earth-chronicles.ru/news/2016-12-16-99404

27. Turner K «Encounter Phenomena Defy «Set Pattern», UFO, 1993, 8,1

28. Turner K, Rice T «Masquerade of Angels», 1994

29. Belimov G. «Proximity with aliens. Secrets of contacts of the 6th kind», 2005 (in Russian)

30. Kipyatkov V.E. «The Behavior of Social Insects», Biology Series, Moscow, Znanie, 1991, 2

31. Mack JE «Abduction: Human Encounters with Aliens», 1994

32. Wittenburg B «Shah Planet Earth», 1997

33. EVER DREAM THIS MAN? – www.thisman.org

34. Hopkins B. «Ethical implications of the UFO abduction phenomenon», 1987

35. Hall R «Silent Killers in Sussex», Richplanet TV Production, 2010 – richplanet.net

36. Corrales S «Argentina: Cattle Mutilations on the Rise» – www.ufodigest.com/article/argentina-cattle-mutilations-on-the-rise

37. UFO Magazine, 1978, 4

38. PEREKRESTOK KENTAVRA (Rostov-on-Don), 8, 2000

39. Long G. Examining the Earthlight theory. The Yakima UFO Microcosm. J. Allen Hynek Center for UFO Studies, 1990

40. More complex patterns, the so-called glyphs or pictograms, are of a completely different nature. They are applied remotely using spherical flying probes. Perhaps the appearance of complex drawings is intended to veil the true reason for the appearance of simple circles and the nature of the activities of UFO pilots

41. By Branton «The Dulce Wars: Underground Alien Bases and the Battle for Planet Earth», 2011, 135 pages

42. Hall R, Keyton D, Gough D «UFO and NATO. «The human mutilation is cover up». – Richplanet TV Production, 2014; Part 1 and 2 – richplanet.net

43. Pratt R.»UFO Danger Zone». Kindle Edition, 1996

44 Azhazha VG «Another Life», 1998 (in Russian)

45. blog.newspapers.com/unsolved-mysteries-george-hodel-and-the-black-dahlia/

46. Keel J. «The Mothman Prophecies». 1975

47. Report No. 13 of Project «Grudge» is missing from the official list of declassified Air Force documents.

48. Von Luedwiger J. «Der Stand der UFO-Forschungen», 1994

49. Ivanov L. LENINSKY PUT (Kostanay), November 22 and 24, 1990 (in Russian)

50. Witkowski B. «UFOs, Abductions and Mutilations», 13.04.2016 – panoffolin.wordpress.com

51. «Victims of the Mysterious Bloodsuckers». – NETKONCA.RU, 15.10.2011

52. Shurinov BA «Paradox of the XX century», 1990,

53. Prokopenko I. «Guests from space. Facts. Evidence. Investigations», 2019 (in Russian),

54. «Kobyakovo Settlement: Monster Hunt» – 7factov.ru/кобяково-городище-охота-на-монстра/

55. Ecker D « The Human Mutilation Factor» www.ufocasebook.com/humanmutilation.html

56. Garsia IZ // «REVISTA UFO» 1993, Sept

57. Covo C, Covo P, da Cunha T. O Caso Guarapiranga. – «Instituto Nacional de Investigação de Fenômenos Aeroespaciais», 2004

58. SEKRETNIE MATERIALI, 9, 2000 (in Russian)

59. Corrales S. «Argentina – Cattle Mutilation In La Pampa»//The Journal of Hispanic Ufology, 7.12. 2006

60. Witkowski B. TBV Investigations Case Number: 548146 – 8.04.2002- Human Mutilations

61. «She bled him.» Murdered rapper Cartwright's family lawyer reveals shocking details of case» Report 23:59, 20 JANUARY 2021 -topspb.tv/news/2021/01/20/ona-ego-obeskrovila-advokat-semi-ubitogo-repera-kartrajta-rasskazal-shokiruyushie-podrobnosti-dela/

62. ИА REGNUM (Krasnoyarsk) https://regnum.ru/news/833449.html (in Russian)

63. https://www.liveinternet.ru/users/4582000/post183461433/

64. Kolchin G «UFO, facts and documents», 1991

65. Ecker D // OHIO UFO NOTEBOOK, Supplement, July 1992

66. Ecker D // UFO Magazine, 1990, 2, 2

67. Ecker D // UFO Magazine, 1989, 4, 3

68. Paulides D «Missing 411. Western United States & Canada : unexplained disappearances of North Americans that have never been solved», 2011

PART 5
IMPACT OF ALIEN CIVILIZATION ON HUMAN SOCIETY

NATURAL ABILITIES AND TECHNOLOGICAL WONDERS

The generally accepted materialistic picture of the world today is clearly not complete. It seems that there are some phenomena due to the natural abilities of insectoid aliens [1], but inexplicable from the standpoint of rationalism. We cannot ignore them, since they are repeated many times in the stories of eyewitnesses. However, we do not undertake to discuss their reliability or mechanism.

Insect aliens have a very special type of bioenergetics. The ability of some Chupacabrus to «breathe flames» is well known. In the 19th century, Jumping Jack repeatedly attacked people, unexpectedly exhaling a powerful stream of burning gases in the face of victims: Polly Adams, Lucy Skels, Jane Alsop, Mary Davis. In 1877, Jack attacked soldiers of the military camp at Aldershot. One was burned with a «thin stream of blue flame» [2-4]. The New Jersey Devil also possessed the ability to «breathe flames». In the forest near Osbornville, one local has seen the Devil exhaling «fire» (1880) [5]. In 1909, Mrs. White was attacked in her front yard while she was hanging laundry. The woman screamed terribly and her husband ran to the rescue. He saw how the Devil, standing over the fallen woman, «breathed the flame» [6, 7]. In the forties of the last century, a similar incident occurred in Provincetown, where the Devil doused a «blue flame» of a teenager [8, 9]. In 1964, a guy named Shrum got lost in the woods and tried to escape from «aliens» in a tree.

At the same time, one of the creatures, trying to remove him from there, blew fire on him [10].

Insectoids of many castes not only vomit fire, they also possess burning limbs. Members of the expedition on the island of Kergalen (1840) noted that the tracks of a strange creature in the snow were covered with an ice crust, which could only appear if the hooves were hot [11]. In 1843 in England, during the period when Jumping Jack appeared there, there were reports of mysterious deaths. Bodies with deep scratches and burns were found in inaccessible places and on country roads [12, 13]. It was written about the Devil's footprints in the snow that they looked as if they had been burned with a hot iron brand [12, 14]. Apparently, not only Chupacabrus, but also representatives of other Alien castes possess burning paws. The scalding paws of the strange «doctor», a member of the upper caste, injured abductee Myrna Hansen when he held her on the table [15]. One of these uninvited visitors burned Irina Yampolskaya, a resident of the Voronezh region, when he touched her cheek.

Perhaps the ability to burn objects and even stones is not a miracle at all, but a natural property of the Aliens. The emitted combustible gases, methane and hydrogen sulfide, are natural metabolites of Aliens. And their ignition can occur by creating an electric discharge, as, for example, in electric rays. However, the Aliens have some special «bioenergetics» different from ours. In Delaware, Mr. Minister saw the Devil emit light like a firefly [4]. American soldier Earl Morrison observed creature shoning or blazing, with a green brilliance» over Da Nang region in Vietnam (1969) [18]. In Hammonton, the Devil ranning across the road glowed eerily (2007) [19]. Devil of Oklahoma City was «covered with something creamy, iridescent» (2015) [20]. In Kharkov, the owners of private houses saw Chupacabrus, who «rushed around the yard and was on fire» (2010) [21]. The glow emanating from Devil body sometimes was mistaken for an arbitrary color changing. Mr. Davidson, near Herne Hill, watched the color of the monster's «suit» change from black to white (1879) [3].

No less strange are the eyes of some insectoids. They burn in the dark with a bright red or green, this glow can change its intensity, fading altogether. Probably, the eyes of dead monsters lose their ability to glow and appear black. An interesting story was told by projectionist Nikolai Brussel in the film «Secret Territories. Monsters. Mysteries of Time» (2013), shown on Russian television. In the 1970s be returning on foot through the forest to his village Shemursha (Chuvashia of USSR), he heard a loud falling sound and found the dead monster fallen from a height on a forest ground. He noticed that his eyes were black [22]. In the photograph of the insectoid's dead head presented by Nikolay Subbotin (see above), the eyes are also black. However, in living insectoids, the color of the eyes constantly changed, which was repeatedly noticed by abductants, as if something flashed inside [15].

It is worth remembering the terrible incident that happened to Philippe Schneider. He received terrible burns in the dungeon, where more than seventy commandos died. It is not at all a fact that the underground monsters, most likely Alien combat units, shot at people with «blasters» or flamethrowers [16]. Apparently, the stories about the blasters and flamethrowers used by the «aliens» were invented later. It is more likely that the people were hit by controlled fireballs. Oddly enough, using glowing spheres to remotely control objects, and even flying in the form of fireballs, may be the real ability of insectoid aliens. In 2002, many people were maimed in India and Nepal, and there were even deaths (2002). People were sure that they are attacked by «fireballs», the size of a soccer ball. However, after the attacks, not only burns remained, but also scratches from the claws [23], as after the attacks of the Chupacabrus.

In the Kuban in 2001, the hostess noticed a luminous ball; it flew out from under the rabbit cages, which, as it turned out later, were bled [24]. In the village of Pustovoitovka, during the attack of Chupacabrus, a luminous ball was captured by a video surveillance camera (2011) [25]. In 2012, a strange case of poultry death was noted at the Velikodolinsky poultry farm, close of Odessa. A resident of Odessa, Valkan (a pseudonym), saw a luminous ball over this area, with a diameter of about 6 m. A resident of central Chile, Jose Ismael Pino, saw a luminous ball, and going to the place of his landing, he stumbled upon Chupacabrus; later, dead animals were found at this place (2000) [26]. In the aforementioned case in Chuvashia, fireballs stubbornly flew over the place where the dead Devil fell. Apparently, it was picked up later [22].

Surprising is not only the ability of these creatures to wrap themselves in some kind of luminous field, but also to turn space. After all, the size of the balls is visually smaller than the animal present there. The ability of insects to surround themselves with fields of unknown origin was discovered by the researcher Viktor Stepanovich Grebennikov (1927-2001). He called this phenomenon the «shape field» (the effect of cavity structures – EPS). This effect also included antigravity phenomena [27, 28]. Interestingly, when taking off using the anti-gravity platform created by the researcher, an outside observer could observe its pilot in the form of a luminous ball. Flights on an anti-gravity platform have not been confirmed by anyone, and so far they can, of course, be considered a funny joke of an eccentric enthusiast. However, not so long ago, the flight of a similar platform was demonstrated in short clips on YouTube [17].

V.S. Grebennikov and his experiments with a flying platform

The possible use of antigravity by the aliens is also evidenced by the often observed vertical takeoff of the Chupacabrus. This is how the Owl Man in England, the Jersey Devils, the Moth in the USA and the Chupacabrus in India took off [19, 29, 70]. Sometimes such a takeoff leaves behind clouds of steam or bubbles. There were also disturbances of the electromagnetic field. In the Ukrainian village of Vovkivtsi, with the approach of the Chupacabrus, there were radio interference, interruptions in the operation of mobile communications, and photo/film shooting problems arose as (2011 – 2012) [30]. Perhaps that is why it is so difficult to get a real picture of Insectoids. For example, for a photographer from Lacey, nothing was displayed at the place where the Devil was (2003) [19].

In some cases, the sudden disappearance or teleportation of living objects allegedly occurred.

In 2015, the Argentine website Vision OVNI [71] reported that in La Pampa, near the road, two Devils were shot at, but they disappeared. Similar cases took place in England and in the Volyn region of the Ukraine (2006), where the unexpected disappearances of a «huge black dog» took place. In the village of Veliky Divlin, two women also assured that they had seen the disappearance of a strange creature: «I was looking at the monster, but he suddenly disappeared. And then he appeared from a different side» (2009) [31, 32].

It can be assumed that such effects are due to the ability to suggest (hypnosis), which the Devils manifest no less than the legendary Bigfoot. Usually Chupacabrus easily hypnotize cats and dogs. In Pervomaevka, Zaigraevsky district of Buryatia, Chupacabrus was seen by a local resident. She described the meeting as follows: «She looked at me, and I was shaking all over, as if I were in an electric chair» (2013) [33]. Or what we perceive as appearance or disappearance may be due to the ability of insectoids to move very quickly. A video taken by an amateur camera in the Indian Kuray of the State of Manipur (2018) captures some kind of creature flying at incredible speed over the crowd, distraught with horrorIn the village of Anuysky, Russian Federation, Chupacabrus very quickly ran past two guys and knocked one of them down with a blow to the chest. Devil's speed was such that the guys noticed a flashing shadow only (2009) [34, 35]. Despite attempts to explain such phenomena from the point of view of modern scientific concepts, there is very ambiguous information about the teleportation ability of quite ordinary termites. The movements of the queen of termites from one chamber of the termite mound to another have been recorded, but not explained in any way; similar abilities appeared in Atta ants [36, 37]. Researchers try uncertainly and helplessly talk about hypothetical «tunnel crossings».

The Aliens have other oddities as well. Aliens were often seen near cemeteries and cattle burial grounds. The first time Jumping Jack was seen by a resident of London near the cemetery (1837). In Clapham Road, an elderly woman saw him in the cemetery (1837), at first it seemed to her that he was a man in a dark cloak. The attack on Jane Alsop (1837) took place in the courtyard of a cottage located near the Hamlet Tower cemetery. Appearing at the Aldershot military base, Jumping Jack very quickly moved to nearby cemetery entrance (1877) [3]. The Czech analogue of Jumping Jack, Perak, (1940 -1944) was most often found near the Olsany cemetery [38]. In 1976, in the Mownen village of Cornwell, Owlman was noticed by two girls when they were playing in the church cemetery [29]. Several people saw Chupacabrus at the cemetery near the Rancho El Sabino farm (1994) [39]. Mothman was noticed by grave diggers near Clindenin (1966) [40].

Ukrainian Chupacabrus elso were seen close of cemeteries: in Podgortsy (2009) [41] and Zheleznodorozhny villages (2012) [42]. Attacks on Raisa Seredyuk in Pyatigory village (2010) [43]; and on Maxim Godonyuk in Stepanovka village (2010) [42, 44] took place close of cemetery [42, 45, 46]. A uniquely interesting observation of Chupacabrus was made in November 2013 at a central city cemetery of Chernigov, [47]. Chupacabrus also appeared from the side of cattle burial grounds. Is it a coincidence or not, but a Strange dwarf was also found at the Kyshtym cemetery [48].

It has been noticed that luminous balls are flying near fresh graves [49]. Such abilities, for example, were possessed by Vampires Obaifo and Lugaru, characters of African folklore, flew in the form of a luminous ball [50]. Chupacabrus are clearly interested in fresh human corpses. Sometimes there were cases of mutilation of fresh bodies in morgues [15, 51]. It is unlikely that they are attracted by the «delicious» smell of rotting organic matter, because Chupacabrus leave many victim carcasses uneaten, they are interested in fresh blood only. Perhaps Aliens are collect genetic material from fresh corpses for cloning. This is just an assumption, but in this regard, numerous and terrible stories about appearance of living «dead» are recalled. There is nothing particularly mysterious about the cloning technique. But

Aliens allegedly manage to grow a whole adult organism «from a test tube» very quickly. It is completely incomprehensible how this is possible without movement and learning at least elementary physiological skills.

There is an aspect of this problem that is even more difficult to approach. According to numerous testimonies of abductees, Aliens possess methods not only for cloning bodies, but also for replacing and «resettlement of human souls». Is there any emotionally significant information being overwritten like from a computer to a floppy disk? Recording, or rather, calling a brain using some kind of technical apparatus, was mentioned more than once in different abduct stories [52, 53, 54].

However, it must be admitted that we are not yet close to solving most of these mysteries. It seems that some official scientific directions, especially physics of fields, time and space, are being artificially hampered. Are only militant retrogrades, fighters against «pseudo-science» to blame? Whatever it was *our fear of foreign technologies, misunderstanding of the meaning of what is happening that allows the Aliens to successfully still, as in the dark times, secretly parasitize on humanity.*

ALIEN TRACES IN CULTURE AND RELIGION

The history of the influence of an alien civilization on humanity goes deep into the depths of centuries. Termitoid Aliens, a biological intelligent species, trampled on our planet long before appearance of *Homo sapiens*. According to our estimates, Aliens appeared 40 – 50 million years earlier than our most ancient ancestor, *Astralopithecine*. American anomalist Charles Fort wrote about «cup tracks», up to 10 centimeters in size, which were repeatedly found on exposed rocks: on a cliff near Lake Como in Italy, not far from St. Peter's Cathedral in Ambleteza, in France, on a rock, known as the Witch Stone, in Invernessshire in Scotland, in County Clare in Ireland, near one of the churches in Norway. Chinese rocks are covered with these prints, in India they also are found. They have even been seen in Antarctica! Like humans, monsters moved on two legs and left strange footprints that looked like those of a bird or a goat [12]. It is clear that the petrified footprints must be of very ancient origin.

Perhaps, even since then, in some underground isolates, giant insectoids have developed their material culture. Grandiose megalithic structures, the remains of ancient roads still remain on the surface of the earth, and some artifacts of this age sometimes fall into the hands of researchers. Mysterious texts, images and maps on disks and flat stones in our time perplex archaeologists. It is a pity that serious specialists do not deal with such findings, but only popularizers of dubious ideas [55]. They put forward hypotheses that go far beyond their own specialty, not caring in the least about how their conclusions correspond to data from other sciences, for example, anthropology and paleontology. After all, it is absurd to attribute traces of a bipedal creature, tens of millions of years old, exclusively to humans. Assumptions about human participation in construction of megaliths or in grazing of dinosaurs are meaningless. In such ancient times, there were no people and no primates in general. At Jurassic end there were only primitive small mammals that had just split into marsupials and placentals. Most likely, all these finds belong to this strange ancient civilization of Alien-termitoids. These masters, apparently, possessed the technique of processing the most accessible material to them, stone. So far, only a few items have been found, and the main infrastructure of the ancient non-human civilization is hidden underground, at the depth of the mines.

But contacts with man and his «domestication» could begin no earlier than the appearance of man himself, and the oldest of the found hominids is no more than 7 million years old. In those days, the first astralopithecines had just descended from the trees and dispersed across the savannah. However, once started, these contacts were never interrupted. It is quite obvious that alien slaves have long been stolen from the human population. It is possible that the capture of archanthropes, their taming and selection could be carried out long before the appearance of modern humans. Some contactees in the Alien Environment saw Anthrop creatures with clearly archaic features [56]. For example, there were individuals completely covered with hair, resembling a bigfoot. However, given that the first hominids were not capable of making even primitive devices such as a scraper, and their domestication did not make sense. It is clear that the «domestic animals» of the Aliens should have been capable of elementary objective activity and of hunting warm-blooded animals. Therefore, domestication could become more productive much later, with the advent of *Homo erectus* . These primitive people were already capable to do elementary work, they left us many artifacts of Acheulean culture. So, more ancient finds, older than 2 million years, are unlikely to belong to the products of human hands.

Thus, the intervention of Alien breeders in Homo sapiens evolution as a species is not excluded, although it was unlikely to be as global as creationists are trying to present it. Apparently, the role of Aliens was limited to trapping and enslaving individual representatives of human tribes. Probably, the breeds of domesticated Anthrops were created by Aliens at different times and for different purposes. It is very possible that the Aliens have been secretly parasitizing on humanity for many thousands of years.

And now, along with those who are significantly modified, like the Grays, there are those who almost do not differ from our contemporary. *And this is the most disturbing aspect of this problem. After all, there may be many such «people» among us.* The Aliens need agents who can quietly infiltrate human society. And such agents, often loaded with implants, zombified by certain methods, really exist. Aliens purposefully train them, often such competent people are considered «magicians», «sorcerers» or «witches». In certain periods of history, they were purposefully hunted by the clergy competing for influence.

A photo portrait of the Alien and
ancient petroglyphs from different parts of the world

Traces of ancient contacts with insectoids remained in legends and in rock paintings. In the folklore of peoples in the southern part of the African continent, the image of a giant praying mantis is preserved. Among the South African San tribe, even the name of the creator god «Kaggen» literally means «Praying Mantis». Images of huge insect-like «horned» creatures can be seen on the petroglyphs of Tanzania, Namibia and South Africa. The age of some finds is estimated at 30-40 thousand years. Huge insects are depicted next to small figures of people. There are stories about the capture of people by strange creatures with characteristic outgrowths on their heads [57]. Similar creatures are depicted on the rocks of the Nazca Desert in Peru and in Canyonlands National Park in Utah.

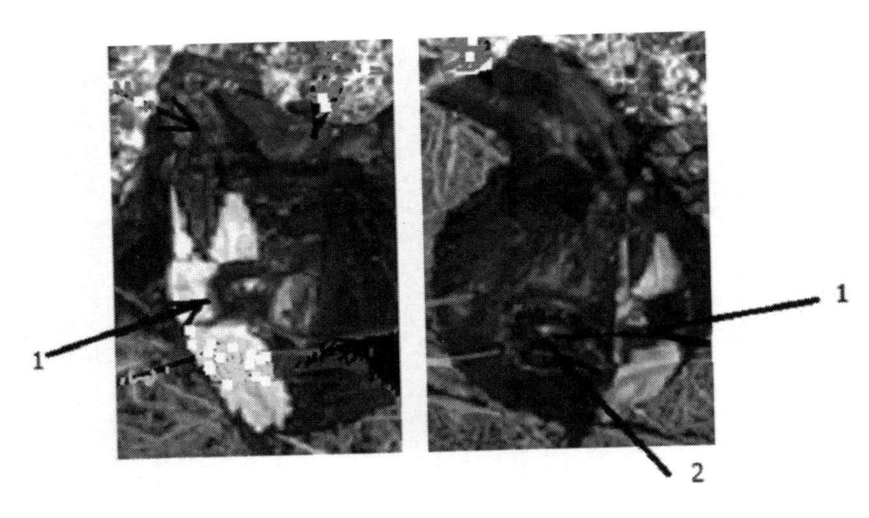

The front (1) and back (2) eyes of the Alien

Creatures very similar to alien insectoids are also described in biblical texts: «... the likeness of four animals was seen, and such was their appearance: their appearance was like that of a man; and each had four faces, and each of them four wings, and their feet were straight, and the soles of their feet were like the sole of a calf's foot, and shone like bright copper, and the hands of men were under their wings... each had two wings that touched one another, and two covered their bodies. And the appearance of these animals was like the appearance of burning coals, like the appearance of lamps...» [58]. A story of Prophet Ezekiel accurately describes insectoid aliens. Upright creatures with double wings covered the body, were just like of legendary Jumping Jack. The forelimbs were under the wings and were clearly not identical to them (therefore, the animals were not vertebrates!). Characteristic feet were similar to a cloven hoof. Eyes were burning. But one detail does not match: each animal had «four faces». It is possible, however, that numerous scribes and interpreters of ancient legends confused eyes with «faces». Perhaps they meant four eyes looking in different directions, like the insectoid in the photographs (it has paired front and paired hind eyes; they are about the same size, so it is not clear which of them are main and which are additional).

Marduk fighting Tiamat

Signs of insectoids can be easily recognized in the human culture.Widely known images of the ancient demigods, Pan and Faun. These characters with goat legs and horns were later transformed into the familiar image of a devil with cloven hooves. The smell of «sulfur» (or rather, hydrogen sulfide and rotting organics, mercaptans), has traditionally been considered the hallmark of evil spirits since ancient times. Such a stench preceded the appearance of an unsympathetic devil visitor. This is how the Shulikuns, the spirits of the northern Slavs, smelled. The Yakuts believed that the «lower world» is a country «with a pungent smell, reminiscent of the smell of an ermine» [60, 62]. The aliens from the underworld traditionally had red flaming eyes.

Everyone knows the legends about the «chud buried in the ground», and, as a rule, these people are considered to be people. However, according to the legends of the Komi-Permyaks, Chud are «eared creatures», their legs end in hooves similar to pigs [68].

However, in many cases, strange footprints of Aliens were perceived as bird tracks. As a rule, monsters have wings and are capable of flight. In ancient Akkadian-Sumerian mythology, there were evil monsters, winged demons, Chaos, Tiamat and Pazuzu, an evil demon with two pairs of wings [59]. Very often, descriptions of monsters resemble giant horned birds, reminiscent of the long-eared owl *Asio otus*.

Ishtar, one of the main goddesses of ancient Sumer, appeared in the form of an owl woman. Later, her Babylonian counterpart Lilith appeared [59]. In the Semitic languages, the word «lilith» also denoted an owl. This ancient demigod of Jewish legends, the bloodsucker, was also winged, hairy, his legs left chicken prints. The ancient Greeks and Romans had a similar demon associated with the owl. The Greeks called her Lamia, and the Romans – Strix (ancient Roman «owl»). Vampires with a consonant name were present in the folklore of Romania, Bulgaria, Albania, Poland and the Ukraine. Both in Western and Eastern Europe, vampires were very much feared. However, bloodsuckers preyed not so much on people as on animals, because the appearance of a vampire was accompanied by a loss of livestock in the vicinity. Animals died from blood loss and concomitant infections [61]. The Latinos on both sides of the Atlantic had a folklore character – Lechuza, also an owl with a female face. This creature attacked from an ambush [61, 63]. The Bulgarian Mratinyak was traditionally depicted as a big scary black chicken with gigantic wings and huge eyes.

Among the Indians in the Pacific and in the Great Plains of America, the Thunderbird, still sits on totem poles. From the eyes of these creatures «lightning flashes», and according to early legends, Thunderbirds were able to steal and kill livestock and people [60, 62, 63].

In the region of Gibraltar lived predatory winged Perithii, with sharp and very hard claws. These monsters attacked people and animals. During one of the Punic Wars, the soldiers of Scipio were allegedly attacked by them. A whole gallery is made up of images of winged Gargoyles. In Lorraine they were called Graulli, in Provence they lived – Gargel, in the British Isles – Grindilow. The predatory winged creature Gargoyya (Gargoyle) attacked animals and passers-by in the vicinity of Rouen [64].

The Indian winged demigod with the consonant name Garuda, a character of Hinduism and Buddhism, was depicted with sparkling golden eyes. Initially, he was also a predator. Among the inhabitants of the Philippines and Southeast Asia, there is a belief in aswangs, flying creatures that combine the signs of a werewolf, vampire and witch. Their very long and sharp tongue, hollow inside, is able to pierce the skin and reach the blood vessels of the victim.

Apparently all these demigods left imprints like of chicken (or forked of goat). According to traces on the ashes, some peoples to identied of evil spirits. The Slavs showed presence of «navi» in this way. The mention of chicken prints of the devil on scattered ashes is in the Talmudic treatise «Brakhot» [65, 66, 67].

It is also alarming that the «owl» and other ornithological demigods most likely symbolize the same Insectoid Alien. The image of the winged monster, as shown above, left a significant mark on the history of art and world folklore. After all, Alitns were often mistaken for huge owls. For example, a very large winged creature, observed in the Mawnan village in Cornwall and called, «Owlman». The devil really looked like a big owl with sharp ears, the size of a man. His eyes were red and shining [29, 70]. In 1976, a similar owl-like creature appeared in El Yunque National Park in Puerto Rico. It was about 1.2 m in size. Its wings were folded and tightly pressed to the back, the lower limbs were strong, and the upper limbs were thin and ended in claws. The monster had glowing red eyes, its mouth was ajar and two large fangs protruded from it [72]. Ten years earlier, similar «owls» were seen near Point Pleasant in West Virginia. They were huge owl-like creatures with huge glowing red eyes. Their large wings were folded behind their backs [73]. Of course, these monsters are not ordinary owls or other birds: under the action of an incomprehensible force, they fly vertically upwards!

Bohemian Mysteries of different years and their symbol

Despite the seeming fantasticness of these images, such creativity can be found in different parts of the world. Sinister monsters reveal a surprising resemblance.

Oddly enough, but the ancient connections of human leaders with the world of Aliens have survived to our time. The place of the annual meetings of high-ranking representatives of government and business is located on a private property in a redwood forest near the city of Monte Rio, California. «Bohemian Grove» belongs to a private club. The most influential people in the world come here to have fun for a couple of weeks. Something is happening here that resembles the ritual of the cults of Moloch, the Canaanites, the Druids, as well as the ancient Egyptian civilizations and Babylon. The main sacred symbol is located on the shore of a small lake. This is a 14-meter cement statue of an OWL, mounted on a steel plinth. Costumed performances and mysteries are played in front of the statue. The costumed priests of the cult address the owl with these words: «Oh you great symbol of all mortal wisdom, Owl of Bohemia... give us your advice!» [69]. Images of the same «owl» can be found, for example, near the Capitol building and on dollar bills.

It is difficult to say how seriously this show is taken by participants of the Bohemian Gatherings, and whether they understand the hidden meaning of the performance. It is evidante this like to ancient traditions to pay blood tribute to vampire god. Apparently, due to similarity of this show with ancient bloody cultsparticipants their partisipants traditionally suspected of committing human sacrifices, organizing blooding orgies and kidnapping children.

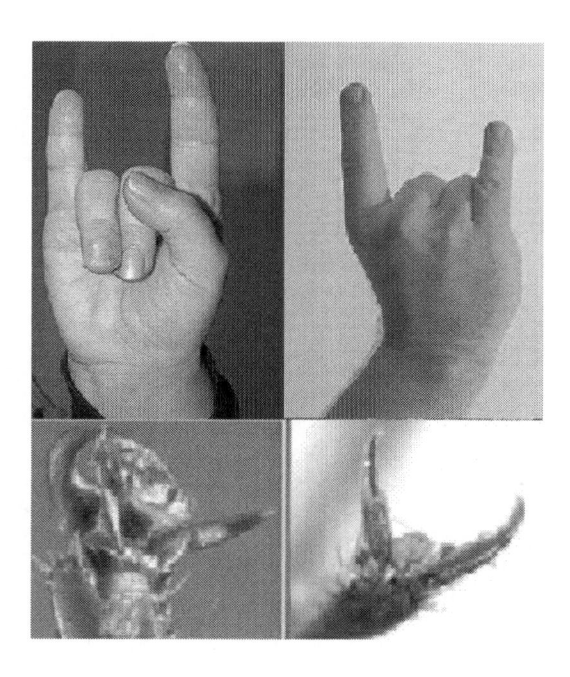

Goad gesture and insect paw

To whom do those in power demonstrate their loyalty? It is not difficult to guess to whom these remnants of ancient cults that have survived to our time are oriented.

The expression «blue blood» used traditionally emphasized kinship with monarchical persons, proximity to the elite of society. It is interesting that, as we saw above, the blood of Aliens has a blue tint. The favorite gesture of the representatives of the world elite, the «goat», it is used by Satanists as a trigger in the psychological programming of victims. Coincidentally or not, the gesture is very reminiscent of an insectoid paw. Probably, with this gesture, an elite demonstrate closeness and subordination to Aliens. Artists of various genres are fallowing, the same gestures and a set of characteristic images can be seen at concerts of famous singers.

It should be noted that not only the image of the Owl was used to personify the power of powerful Aliens. In the early 1930s, images of winged bats became popular. Batman is the personification of a certain secret group of people from the dungeons, dressed in special costumes. They supposedly can fly, move at high speed, suddenly disappear. In 1930s, Batman becomes the hero of mass comics, such as «Spider-Man», which also in a certain way reflects the image of Alien.

A crowd near Fatima village on October 13, 1917

The objective existence of a parallel civilization and its influence on many masses of people have taken place throughout the history of mankind is obvious. First of all, this found manifestation in the existence of cults and even world religions. The apparitions of saints, prophets, are supported by the demonstration of «miracles» from the technological arsenal of Aliens, masquerade and deception of gullible and naive people are used [53, 54, 75, 76].

MENTAL INTERVENTIONS

The mechanisms of the influence of an Alien civilization on human society were studied by Dr. Carla Turner (1947 – 1996), a serious researcher, professor of psychology at the University of North Texas. Her interest in the topic of abductions was not accidental. She herself, members of her family became victims of abductions. Noticing strange phenomena, she tried to figure out what was happening. In the future, she created an entire school in which, together with her like-minded people, she tried to help people with psychological problems caused by contact. Carla Turner had a predecessor, psychiatrist Dr. John Mack, who worked with kidnapping victims for forty years. Many other researchers wrote about this phenomenon, but they treated abductions as a rare phenomenon important for establishing contact with an alien civilization and were of the opinion that such contacts help people in mental and cultural development. However, Turner had a different opinion and emphasized the exceptionally hostile nature of the Alien Civilization. In one of her books, Turner wrote:

«I want people to know the truth about how hypocritical their fellow human beings really are. I want them to know that the great and wonderful aliens are actually like demons, which are not supernatural at all, but they are the same physical beings as we are, only using high technology. I want people to stop being so gullible and start asking the right questions... I think of all the innocent victims, especially the children who daily resist the false power».

According to Carla Turner, contactees are under intense psychological pressure. This is intimidation, physical impact, and suggestion to the contactee of the idea that the kidnappers are acting for his own good, thereby justifying his fear, pain and injury. Most of the abducted are completely zombified and unable to adequately assess what is happening. Many contactees were instilled with the ideas of their «chosenness» for the sake of some high goal, they were promised individual salvation in the coming Apocalypse. Most often Grays or Anthropes, who are not much different from ordinary people worked with abductees. They tried to evoke feelings of devotion, gratitude, or affection towards themselves. Each time the suggestion was accompanied by reinforcement in the form of scenes designed for the intellectual level of the kidnapped. As a rule, among the «chosen ones», it is low. Eleven-year-old Pat saw «angels» and «Jesus» (1954), eight-year-old Teddy saw a man dressed in a bright green musketeer jacket embroidered with gold (1955) [53, 54]. And after almost thirty-five years, when people had ideas about spaceships and astronauts, the contactees saw the Antrops dressed in something like an astronaut's overalls.

As a rule, contactees are instilled with an idea that brothers in mind arrived from other planets or even from other star systems. Contactees were arranged demonstration tours there. They are told about the existence of omnipotent extraterrestrial civilizations, allegedly controlled by representatives of a certain «Intergalactic Council». Trusting contactees are told that many civilizations were founded by «humanoids» (not at all embarrassed that this is impossible either from a historical or biological point of view). At the same time, any role of Giant Insectoid Aliens is not advertised. They were only occasionally present in some scenes where they were clearly directors, as was in Linda Parker [77], or Theodore Rice abductions [53, 54]. (May be much more often these cryptids are associated with reptiles, although there is no reason).

In the stories of contactees and abductees, one can often find interesting, as if confirming, coincidences. There is an opinion among researchers of contacts that they testify to the reality of the experienced experience. However, repetitive details can only indicate one thing: all these stories are broadcast from one center.

Their purpose is to further disseminate the ideas of the extraterrestrial nature of aliens through agents of influence that have undergone processing to other members of human society.

It is doubtful, however, that everything is so blissful. Perhaps this is a pseudo-memory, shielding something else. Although description of alien worlds may look fascinating, the real situation may be completely different. There is evidence that among the Aliens, in concentration camps, insect guards use human labor. It is not known whether these mines are located on Earth or somewhere outside it. Another abductee said that he had been in isolates, where people live and work, who, like in a concentration camp, are guarded by aliens, not allowing them to move freely. A little girl, one of those who were lucky enough to return «from there», explained that children are raised there, because adults who got there go crazy [91, 92].

Many abductees have described something like a machine-assisted «brain scan,» in which headphones were put on the subject and electrodes were attached to the forehead. It is proved that some contactees were implanted with small high-tech structures made of metal and other materials. Dr. Roger Krevin Leir (March 20, 1935 – March 14, 2014), an assistant professor at the California College of Medicine and a respected surgeon with an impeccable reputation, removed many such objects from the bodies of victims [78]. These items were implanted in different parts of the body, but most often they were placed in the nasal or ear sinuses of the skull. Perhaps their purpose is to receive and transmit information, stimulate or block any motor activity, etc.

Even with the help of hypnotic regression, it is not always possible to understand the real picture of what happened. Carla Turner believed that in the Alien Environment, the abductees were subjected to powerful psychological influences and hypnosis. According to Turner, people's conscious memories of abductions are reliably blocked. Moreover, for reliability, they are screened with «false» episodes. In some cases, several more layers were hidden under one layer of pseudo-memories, their parallel chains of events existed. Alien technologies allow you to create any illusion in a person under control. Turner mentioned, for example, a «virtual reality scenario,» an externally induced vision accompanied by full sensory input. Subjectively for a person who is «inside» the illusion, it is practically indistinguishable from objective reality. For an outside observer, at this moment the person is immobilized and is inside a bluish luminous «cocoon» or fog [53]. In general, this resembles the described method of Alien communication, described in REPORT-96 with the help of dynamic spatial images recreated by their magnetosensory organs.

Therefore, psychologists sometimes find it difficult to separate the real memories of contactees from what was artificially brought into their memory. Although some experts believe that the information obtained through hypnotic regression is unreliable, it is clear that, while in a state of trance, the witness cannot consciously lie. However, he can broadcast someone else's lies. And that's what Carla Turner and hypnologist Barbara Bartolik were trying to figure out.

In any case, the hypnotic regression is still almost unique method for studying this strange parasitic civilization «inside.» Aliens and Anthropes manipulate the perception and psyche of a person, using technologies far beyond the limits of human capabilities. The result of interference in the psyche and physiology is complete control over a person's behavior, information about his whereabouts, mode of action and even thoughts. After abductions, which can begin as early as childhood, a person is under the constant and strict control of Aliens for the rest of his life.

Spreading stories through their agents, processed, is the basis of planting cults. It allows you to subordinate the consciousness of people to the interests of another biological species. A well-known phenomenon in the Portuguese village of Fatima, which can no doubt be attributed to ufological events (1917) [79], was very indicative.

According to Dr. Carla Turner, all of the abductees have demonstrated above average psychic abilities (although it is not clear if this is a cause or a consequence of the abductions). Psychiatrists consider the ability to hear «voices» to be a manifestation of schizophrenia, and, according to them, about 15 million people worldwide live with it. According to a joint study by two Manchester universities, one in twelfth children in Britain has heard voices at least once in their lives. In November 1990, the First British Conference took place, organized by members of the Manchester group of «hearing voices» [82 – 84]. However, «Voices», telepathic and sound contacts with persons who are physically absent, are an objective and fairly common phenomenon. This phenomenon has been known for a long time [85]. As a rule, miraculous «radios» included spiritual questions, warnings of future disasters, teaching, demonstration of symbols, and transmission of information from the fields of mathematics, physics, politics, and nature. It should be noted that often such messages did not demonstrate a high level of intelligence and even contained targeted misinformation [52, 54]. Voices teach, warn, guide, connect with other contactees through audio contact. It is believed that such «seers» were Publius Cornelius Scipio (236–183 BC), the Frenchwoman Jeanne, who later became the famous heroine Jean d'Arc [86, 87].

Helena Blavatsky (1831 – 1891) also heard «voices» from childhood. She wrote: «The recipient of the Teaching is not accidental, and the language is chosen according to need.» She allegedly possessed a whole range of knowledge received from unknown «teachers» [88]. Interestingly, when communicating directly with Aliens or Antrops, contactees almost always received messages «telepathically». It is possible that signal transmission occurred at the level of the auditory nerve. It is still difficult to understand whether the nerve is excited by vibrations of an external electromagnetic field, or whether the transmission is carried out directly to the auditory centers of the brain. Judy Dorathy, having been on board with the Grays, heard the sounds of a «meowing language similar to Chinese». However, when the speech was addressed to her, she perceived it as being delivered in her native English [15]. The Strangers spoke to the Contactee from Tbilisi in Georgian, and to the Russian biologist in Russian [96].

Communication with the help of implants should not be surprising. It is no more «wonderful» than our radio or mobile phone. At the end of 1987, Polly heard the «voice» of a «Russian professor» living «near Kyiv», he tried to communicate in French. «Golos» called himself «Evek». According to Carla Turner's informants, the Aliens use implanted chips to create a whole army of controlled agents [54]. Connection with the «world of spirits» can have a key influence on political decision-making, the formation of state ideology. This was written by Nicholas Goodrick Clark, a British scholar of religion, director of the Center for Western Esoteric Studies (EXESESO) at the School of Humanities and Social Studies at the University of Exeter [89]. In 1973, the Raelian religious movement appeared in Canada and has built a UFO recreation park near Montreal. Claude Vorilhon, a former sports reporter was a founder of this sect, but he has changed his name to Rael. Raelians claimed that life on Earth was created 25,000 years ago from DNA molecules imported by «Aliens» [95]. All this is complete nonsense, but intention to sow this idea is obvious.

From a trivial point of view, at present, one of the goals of the Aliens may be to stop the exponential development of mankind and a significant (but not complete!) reduction of human population. Perhaps this is precisely what the aggressive propaganda of non-traditional sexual relations is used for, leading to a decrease in the birth rate and the spread of eschatological teachings among adolescents and adults. Over the past half century, cases of the emergence of suicide sects have become more frequent in the world. In 1978, about a thousand members of the Peoples Temple in Jonestown, Guyana died from cyanide poisoning. In 1993, in Kyiv, adherents of the White Brotherhood sect intended to burn themselves near the Church of St. Sophia of Kyiv. From 1994 to 1997, members of the «Order of the Temple of the Sun» died in Quebec. At the end of March 1997, 40 Heaven's Gate sectarians died at the Santa Fe ranch in California. A similar incident occurred in January 1998 in Spain, where the police managed to prevent a suicide of three dozen people in Teide National Park. In mid-March 2000, 778 members of the Ten Commandments Revival Movement in Uganda died. For a while, a similar community existed in Toronto, Canada.

Who leads the human herd to slaughter? One of the Turner abductees, Amy (a pseudonym), reported that in January 1994 she attended Omega meetings in the Dallas area. Having been at this meeting, abducted Amy learned that they were led by four tall, thin and pale people of indeterminate sex, tightly controlled emotionally. They traveled around the country, recruiting groups of people. Their cult requirements were rigid and rather strange.

Adepts were required to break all human affection ties, live outside the family and pass all decisions to the «senior». These teachers talked about exactly what the «voices» taught Amy from childhood. They explained that Jesus was a soldier who had to lead people in some direction [54, 97]. Amy, Ted Rice, and other contactees watched Aliens captured peaple groups in UFOs. There were those who had already been in Aliens Enviroment, but there were also completely random people for whom what was happening seemed unusual and terrible [53, 54].

Perhaps, in our society, among us live not only our compatriots with implants, but also Anthropes themselves, who are not much different from us [53]. These are the people to whose birth Aliens are most directly related. One of the authors of this book is aware of at least three such cases. There are even a couple of photographs of such people. About themselves, they reported that they were «never born», but are among us not of their own free will. For ethical reasons, we cannot publish photos of these unfortunate people.

According to some abductants, insectoids and anthropoids are in perfect harmony. This is what Vladimir Petrovich Stoyan, a resident of the village of Ilyinka, Krasnoperekopsky district, born in 1975, told Anton Anfalov. One day, he was moved to an isolate, which he considered to be another planet. He was accompanied by an anthropoid of a completely human appearance. However, other creatures were also present, which, according to the description of the witness, were typical Aliens. There was something like processes on their heads, their skin was «scaly», the eyes were huge, the teeth were sticking out of «mouths», claws were on their «hands». They were wearing strange black shiny clothes and walked with a shuffling, waddling gait from foot to foot [94]. As the abductee realized, anthropoids and insectoids have a peaceful, if wary, relationship.

One day, a naturopathic doctor from the Voronezh region of Russia, Irina L., treated a slightly strange lame woman. She complained about life, claiming that she was thrown from another planet for some sins. Irina was sure that the mentally ill patient was fooling her. Realizing that they did not believe her, the woman said that her words could be confirmed by her «BROTHER». Soon the brother appeared, and he was not alone.

They were two insectoids in «cloaks with hoods.» They had greenish eyes without pupils, which turned red from time to time. A useless conversation took place between the participants, the details of which the shocked Irina did not remember well. However, there is a very revealing moment in this story: an anthropoid woman called an insectoid «brother». (Such an attitude towards Aliens among abductees, and possibly Anthropoids, is typical. However, it is not at all a fact that Aliens reciprocate them).

MILITARY INQUISITION

One of the most interesting things in Carla Turner's books was the revelation of the direct involvement of the military in the contact. Since 1954, there has been widespread surveillance of contactees, the military detained them, subjected them to interrogations and intimidation, used a «lie detector», and administered tranquilizers. Moreover, even to children who have become victims of abductions. All this was justified by the interests of «national security» [54]. The «Department of the Navy», was engaged in surveillance and monitoring of abducted people. However, security services was instructed not to take on any role other than supervising the kidnappings. Unmarked black military helicopters circled over their houses. Their task was to monitor the activity of other people's «plates». There are even cases when the appearance of such helicopters prevented abductions [15, 52, 54]. Abductee Anita (pseudonym), who lived north of Houston, fearing further unnecessary problems, quite deliberately refused to answer questions from the military and pretended to know nothing (1977) However, many contactees detained by the military unconsciously fell into a stupor, clearly induced from the outside. Carla Turner cites many such cases. Turner's husband, Casey, was also flown by an F-150 helicopter to a remote underground military base (1988). Casey recalled moving through a narrow tunnel filled with dusty, bulky equipment. According to Casey, there, at the base, he saw only the American military. The guard escorted Casey through a fairly large hall, where some of his acquaintances and strangers were sitting in suspense, all were in an «inadequate state». They all just sat there in a daze, as if they were drugged. During the interrogation, a major in his fifties in a green uniform with a black stripe on his arm, without a name plate, sought truthful testimony, threatening to himself or members of his family. They tried to find out from him what exactly he knew about his abduction. But they could not get anything from Casey, he himself did not understand the reason for his inhibited state [52].

Almost all contactees from the Turner group fell into the same inhibited state before interrogation by the military. She explained this effect by remote influence on the part of the Aliens, who were trying to avoid information leaks [52, 54].

All this fuss with contactees and ufologists studying the problem speaks of the lack of control of the situation on the part of the military and their panic. Otherwise, what was the point of torturing the unfortunate abductees? Follow their movements, and at night control the territory adjacent to their houses from helicopters? The military was very interested in what exactly contactees see in Alien Environment. Naturally, within the framework of the counter-game characteristic of intelligence, they wanted to enter into a dialogue with unknown forces. It seems indisputable that for many years American intelligence officers from the Air Force and the NSA tried to establish a profitable contact for themselves, apparently, they themselves had little understanding of what, in fact, they were dealing with. Perhaps it was for this reason that studies were carried out on the influence of extraneous forces on the human psyche. Mid 1950s McGill University in Canada conducted mind control research under the direction of Dr. D. Ewan Cameron. His lab conducted these experiments under US CIA program called «MKULTRA». This project was closed in 1964, but secret research aimed at influencing human memory continued in the United States and Canada as part of «ARTICHOKE» and «MONARCH» projects.

According to some abductee information, it is possible to recreate a history of interactions between Aliens (or their symbionts) and human military. If in the mid-50s the military were helpless, apparently, by the end of the 1980s, the tasks of the military inquisition were partially completed. Most likely, contact was established and cooperation made it possible to borrow some technologies. In 1989, Angie was took out of her house at night to a military base. She was showed some «Alien» technology in action and informed that she, like several other contactees, had been selected to participate in a mind control project. The base where Angie ended up was in Northern Arizona. A man in civilian clothes (most likely from the NSA) was apparently boss of military. He told her that such underground bases are located around the world, including bases in New Mexico, at the North Pole and in Africa. Their operation was named like «HIGH JUMP» or «HIGH SHELF». This civilian chief claimed that both the military and the anthropes were subordinate to him. When Angie asked if the Aliens («large gray») obey him, the man replied evasively, saying that this would also happen, but «gradually».

It seems that Antrops helped American military to a large extent in mastering some methods. In any case, they were repeatedly seen at military bases by contactees from the group of Carla Turner: Lisa, Jane, Amy, Angie and Pat (all pseudonyms) [52, 54]. Some anthropes helped the military find and inactivate implants in abductees bodies.

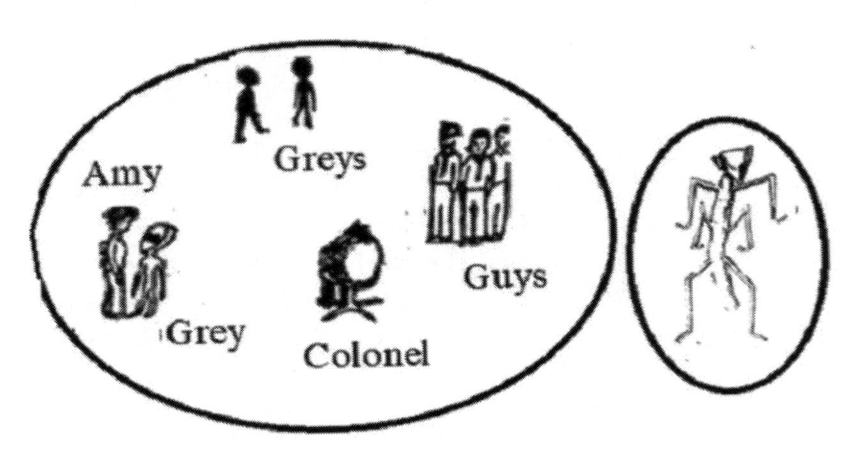

Members of the «Resistance» at an underground military base (left) and «Hostile termite mound» (right)

In November 1992, Amy (pseudonym) was taken to an underground military base. There, in the spacious hall, she saw a lot of Antrops, little Grays and three guys of almost ordinary human appearance. They differed from ordinary people only in their large black eyes. The Grays were wearing masks, which she was told made breathing easier. The little Gray Being explained to Amy that he was human, but of a «different race». This human race was allegedly bred by some other aliens, «hostile» to people. There was also an American army colonel present. He had very piercing hypnotist eyes and gray hair. Amy felt like he was in charge here. She portrayed this scene: a dozen anthropes, three strange «guys» and a completely human military man, located in a chair. Amy explained that there was a special military group that, along with a «resistance» group, had formed among the Antrops. Amy had two implants removed in the medical office, one from her ear and one from her spine. However, the third implant remained in her cerebellum. The group, unfortunately, did not have an equipment to safely remove it [54]. The task of the group included not only the detection and inactivation of «Alien implants», but also the installation of their own, with a specific code.

This evidence is not the only one. Other members of the group of abductees with whom Carla Turner worked repeatedly spoke about such cooperation. Among the methods that the Antrops apparently shared with the US military were methods of «brain scanning» using headphones and an electrode placed on the forehead [52, 54].

It is also noteworthy that at the military bases, the witnesses saw mainly the Grays or Antrops, who agreed to voluntary cooperation. However, the Italian ufologist Valereo Loce, a professor at the University of Corrado Malanga, also wrote about the presence at military bases in Aliens. In his book UFO in the Head, the information received from the contactee Valereo was analyzed. In addition to the American and French military, there were small and large Antrops, the contactee also saw Aliens (creatures with an elongated muzzle and «tail», strange huge grasshoppers). The military spoke French to Valereo, and English to the insectoids. There were also Israeli soldiers [93].

REASONS for COVERUP

When carefully studying the problem of the Alien Civilization, one can notice that throughout the entire 20th and current centuries, the mane task of competent authorities was COVERUP of information. Conspiracy theorists like to write about some kind of secret arrangements allegedly existing between the US government and the «world elite» on the one hand, and Aliens on the other. However, the reality of such an agreement is doubtful. Aliens have always used their light disks to monitor places of production, storage and transportation of nuclear charges, places of testing of new military space technology. They do not need permission to build underground bases, as well as to capture people and livestock for their needs in any region of the world [105].

In 1940th American authorities were extremely concerned about UFO sightings. The High Commissions (MJ-12), specialized commissions: Robertson (1952), Condon (1966) created, additional civil (ANB) and military intelligence agencies were created, the divisions of which jealously divided roles among themselves [99-101]. All this history of American establishment has not direct relation to the real manifestation and significance of a phenomenon, in which unknown flying objects are just a tip of an iceberg. Now, from time to time, another portion of declassified documents appears. Russian ufologists love to flaunt information about these American programs, although it does not carry any real information about the problem of our neighboring civilization. As a rule, most of these projects and programs dealt mainly with external manifestations of the activity of the Alien Civilization. It is impossible to understand its nature based only on the reverse engineering of downed saucers, dead and often severely destroyed bodies of UFO pilots (usually anthropoids), and even more so, focusing on the false testimony of the Grays – it is impossible.

However, the international establishment seemed to be interested in maintaining the COVERUP regime. The Bilderberger group, created in 1952, according to some conspiracy theorists, was carrying out intergovernmental oversight under «ALTERNATIVE 3th» program (ways of «peaceful coexistence» with aliens) [104,106]. Paul Stonehill, director of Russian Ufology Research Center in the USA, wrote that during 1956 heads of the secret services of the USSR, the USA, France and the Great Britain met repeatedly in Geneva, where they developed a common strategy. There is every reason to believe that it was these intelligence services that developed the main approaches to UFO witnesses, including pressure on eyewitnesses by warnings, threats, and even physical violence. It was proposed to invent false explanations for strange phenomena or completely hush them up in the media. It was considered very important to introduce people into amateur UFO groups, to direct their activities.

The fact of secrecy using does not mean an explanation of its causes. On November 11, 1975, a public relations instruction came out of the office of the US Air Force Security Secretary to avoid leaking any scary information. Even earlier, in the proposals of the Condon Commission, it was supposed to avoid informing the population about the danger in any way (1953) [101]. This was due possibility of world panic among the population. However, this is absurd. The presence of a powerful non-human civilization is hardly capable of capturing the imagination of those who have been brought up in a religious environment since childhood, and there are many such people on Earth. Catholics, Orthodox, Muslims and Buddhists are unlikely to be confused by the real existence of powerful gods or demons. It hardly makes sense to hide the existence of Grays, Nordics or Dwarves, especially since their images are replicated in modern folklore and can hardly capture the imagination. True, people who are close to biology or medicine understand that if these creatures are not mechanical dolls or «biorobots», then most likely the products of natural earth evolution, thay could not appear outside the Earth.

During his election campaign in 1976, Jim Carter promised voters that if he became president, he would make available to the public all the information about UFOs in the country, provided that the declassification of this information would not harm the interests of the country. However, this did not happen.

Back in 1952, US President G. Truman informed the leadership of the USSR and NATO countries about some events and details related to UFOs [102]. Perhaps this happened even earlier [103, 104]. This means that this information could not harm the United States, however, for some reason it had to be hidden from a public without fail.

As a rule, the COVERAGE has always been justified by the interests of the military departments and the military industry. After all, mastering the progressive technologies of the aliens would definitely give a lot of military advantages. Some of them were described by Corso [102]. However, on the one hand, no contacts helped the German «contactees» win in the Second World War. In downed UFOs, of course, you can peep some «little things» such as contact lenses, Velcro locks, and even microwave engines and microprocessors. However, practice shows that such inventions very quickly migrate to industrial companies that, in the pursuit of profit, distribute high-tech products around the world.

Antrope bodies recovered from crashed UFOs have been studied as part of the American project «Guests» (1953), and on February 20, 1954 they were allegedly examined by President D. Eisenhower in an underground complex in Indian Springs in Nevada. However, in the summer of 1954, American colonel was interrogating 11-year-old Pat, who had been in the hands of the Antrops. He was not satisfied with her story about the gray men she saw in Alien Wednesday. Therefore, the little girl was given an injection, after which she could not lie [54]. So the US military was interested in SOMETHING else? In another case, a military, it seems, the same colonel, strictly asked Lisa (a pseudonym) what creatures she saw in an Alien Environment. She saw four species, but the military was interested in on only, may be Aliens (insectoids) [54]. A retired military man who served fifteen years in Rapid Response Team 58, the «Black ops group» in NATO, said at UFO crash sites he saw the bodies of UFO pilots many times, of which there were three types. The creatures of one of them are extremely strong and powerful, they «looked like devils and could simply tear a person apart» [80]. Surely, we are talking about the Aliens. So it looks like the COVERUP is primarily about Insectoid Aliens. In a telephone interview given to Linda Howe on (date), Robert Cooper claimed that information about Aliens (insectoids) was marked «orange», which classified this type of alien as «special». In this case, absolute non-intervention was assumed» (1989)[15].

Especially carefully concealed not only the very fact of the existence of Insectoids (Aliens), but also their habitat (in the underground cavities of the Earth), but also their physiology and natural needs (in the bloody product), and, especially, everything related to their natural abilities and technologies based on operating fields and space.

There are reasons to believe that, in the mid-1940s, Alien insectoids were already captured by American army. Dr. Robert E. Sarbacher served in the Pentagon where he interacted with employees of Vannever Bush, Robert Oppenheimer and John von Neumann (members of the so-called MF-12 group). In a letter to William Steinman dated November 29, 1983, he wrote that he had the impression that «extraterresrial» bodies were like some terrestrial insects. Another indirect confirmation may be the term «Alienocyanin» used in REPORT-96, it is clearly English. Cell descriptions made in Dr. Burish's report corresponded just to insectoid tissues. In addition, in American sources there are references to «hybrids of insects and humans» (which is impossible in principle!) (see Part 1) designed to confuse readers.

Although, it is fewt known about insectoids moving into UFOs, there is a suspicion that a sphere fallen in April 1989 near Santa Rosalia in Veracruz state of Mexice was manned by insectoids, an very unpleasant smell of theit bodies testifies to this. The wreckage and bodies were taken to the United States. Probably, an accident also happened in January 1996 in Brazil near Varginha, from where the autopsy materials and live insectoids were taken by American researchers [107,108].

All the attention of ufologists was directed to flying objects and their pilots, while the hidden aspect of this problem was little known. Most likely, bodies and living specimens of insectoids had to be obtained underground. The captures of one or two Alien bodies could not give a complete picture. And as we can see, research work had to be carried out *in vivo*, moreover, on a large group of creatures from different castes of Aliens (see Part 1).

It is very likely that back in the 1920s and 1930s, the Chekists of Soviet Russia, having taken the baton from the Tibetan monks through their agent Yakov Blumkin and the scientist Alexander Barchenko, took underground anomalous phenomena very seriously. The same ideas about underground inhabitants were picked up by researchers from Germany. After the Second World War, this problem naturally went to the winners of the US-UK coalition as a legacy from the 3rd Reich. It became obvious to them that something was nesting in the bowels of the earth. And this involved extensive search work in the dungeons [74]. Ray Palmer and Richard Shaver reported that some of the NORAD underground facilities in Colorado were being built in place of pre-existing cave systems, writing as early as the mid-1940s about the government's search for a system of ancient underground caves and tunnels. And in the 1950s, the construction of underground bases began. Similar work was carried out in the USSR (see Part 2).

In the mid-1970s, the problem of livestock mutilation arose and spread to pastoral areas in North and South America and others continents. However according to Philip Corso, since 1951 the US authorities have known about these phenomena and teirs connection with UFOs [102]. Chairman of «Long Island UFO Network» John Ford wrote that COVERUP it was the mane aime of federal government (1975). It is unlikely that among American specialists there were no biologists, biochemists or doctors who would not explain to military that predators cannot be «extraterrestrial» if they are eating the blood of terrestrial animals in any way. Probably, knowing the truth, they still tried to hide it from the citizens by any way. In 1983, after Linda Howe's «Alien Harvest», aired on KMGH-TV-7 in Denver, Colorado, they attempted to discredit it. FBI-affiliated Joe Nickell, Kenneth Rommel, D. John M. King (Cornell University veterinarian). The cover operation was carried out by AFOSI agent [110] Master Sergeant Richard Doughty, from Kirtland Air Force Base in New Mexico. In a personal conversation, Linda Howe was told that she came too close to what the authorities would like hide.

Cases of mutilation of people were hidden especially carefully. However, there was close international cooperation elso. In 1959, after death of Dyatlov group in the Northern Urals, the Soviet leadership held secret consultations with NATO leaders, where 4 reports were received on this case. Apparently, after these events associated with the unsuccessful cleansing, the Soviet leadership began to allow foreign specialists from the «Black ops group» into their territory. NATO had better equipment and more experience. In addition to Russia and the USA, these groups worked in Great Britain, Ireland, Scotland, Spain, Germany, Australia and Yugoslavia. These operations were always led by American military, American scientists taken all materials from the places of human mutations [80].

It should be noted that in the 1990s, UFO raids on livestock farms decreased significantly. This is probably due to the development of laser guns capable of shooting down flying objects. But it was at this time Chupacabras rised, may be for compensating of food foraging for nests.

No wonder the US military and explorers tried to penetrate the depths of the earth (in the USSR they did the same, but even more secretly, and sometimes they simply undermined the portals, preventing Alien evil spirits from getting out). Perhaps the military sought to master the underground spaces, displacing the inhabitants of the nests from there? In 1989, Beth (a pseudonym) was arrested by military and taken to an underground military base in an unremarkable desert area at the foot of a hill. Behind an inconspicuous wooden door were technological rooms, where people in military uniform and civilian clothes worked at desks. At some point, two employees in anti-radiation suits appeared, the alarm was raised, and those present fled [54]. Perhaps there was a risk of explosion?

The proximity to the nests was very dangerous, although the US military struggled to get into them. It is believed that Project Beta was aimed at destroying one of these nests near the Dulce of New Mexico. According to a local resident, military Huey helicopters, unmarked, filled with military men armed with rockets and M-1 (or – M16) rifles, often circled near mounting Archaluta Mesa. They did not have any rank indications, only shoulder patches, and the helicopters had no identification marks other than numbers. The helicopters were part of a project called «BLUE LIGHT», they were from Fort Carson, Colorado. Sometimes they accompanied an UFO taking off [109]. In 1979, when trying to build underground buildings in the immediate vicinity of the Alien nest near the town of Dulce, there was a shootout, which was told by former military builder Philippe Schneider. Many commandos were killed there. He believed that the combat situation arose by chance, but this is unlikely [16]. Most likely the military tried to get into the nest.

In 1980, Dr. Paul Bennewitz, an noble American physicist, former president of Space Shuttle instrumentation company «Thunder Scientific Corporation», located an underground Alien facility near Dulce. Bennewitz was able to determine that the underground cavities were not only under Archuleta Mesa mount, but stretched in all directions. After four years of UFO reserches, Bennewitz got silenced, because his problems with intelligence service began. He was subjected to increased pressure from Kirtland Air Force Base command and from AFOSI employees. He was followed by agents of the «Office of Secret Investigations», which he believed belonged to NSA [109]. So why did Bennewitz so annoy the secret services? He believed information about the nest near Dulsa had a secret level higher than President's. In 1988, AFOSI agent Doughty wrote that Paul Bennewitz was under investigation. Later ufologist Penny Harper reported in the January 1990 issue of Whole LIFE TIMES that Paul Bennewitz was sent to the New Mexico State Hospital for the Insane, where he received electroshock therapy, his whereabouts since then unknown [110]. In 1996, Carla Turner died of transient cancer. Until last days, she was sure that she had been exposed to some kind of malignant radiation from some military services. She was also very interested in underground military bases location of and their interaction with Aliens [54].

Why world authorities and intelligences are so sensitive to question of Alien-incectoids and underground aspect of their life? A hint of some serious danger emanating from them is contained in REPORT-96. Perhaps for this reason, research on Aliens abruptly ceased, and former employees of the Sechenov Institute are still afraid to speak on this topic. Is the danger real and what exactly is it? Perhaps the mastery of physical laws that are beyond the scope of our modern science is the biggest secret the Aliens have. And this is not only what world powers would like to get in secret from each other. Perhaps, with these technologies, we would pose a threat to Aliens? Until now, ufologists are arguing about the essence of the Philadelphia experiment (28 Oct 1943). Perhaps some of these technologies are trying to master now. Some information about was considered in Internet in 2008 [111]. However, we can neither verify nor evaluate it, so have not an answer to this question.

NOTES to Part 5

1. Agon E «Real Monsters», 2020 (in Russian)
2. Haining P «The legend and crimes of Spring-heeled Jack», 1977;
3. Dash M, «Spring-heeled Jack: To Victorian bugaboo from Suburban Ghost»// Fortean Studies, 1996, 4, 1-125;
4. Begg P «The Terror of London», 1981
5. McCloy JF, Miller R, »Phantom of Pines: More Tales of Jersey Devil», 1988
6. Moran M, Sceurman M, »Weird N.J.: Your Travel Guide to New Jersey's Local Legends and Best Kept Secrets», 2004;
7. »In 1909, the Jersey Devil was sighted in Gloucester», CNBNewsnet, Feb 2007, https://www.gloucestercitynews.net
8. Smyth M, «Toronto Star» 29.10.1988
9. Cahill RE, «New England's Mad and Mysterious Men», 1984
10. «1964 Alien Encounter in California», COSMOSONLINE.RU 3.02.2014
11. Ross JC «A Voyage of Discovery and Research in the Southern and Antarctic Regions, during the Years 1839-43»,Vol1
12. Fort CH «Lo!», 1931, Reprinted 1965 – http://www.resologist.net,
13. Gershtein M «SECRETS OF THE XX CENTURY», 2010, 21
14. «Spring-heeled Jack in India», FORTEAN TIMES , 1996, 91, October, 20

15. Howe L «An Alien Harvest. Further Evidence Linking Animal Mutilations and Human Abductions to Alien Life Forms», 1989

16. Schneider Ph – Speech «Aliens & Underground Bases» – November 1995

17. Gravity Flight – https://www.youtube.com/watch?v=1JeeaZlYonc&t=43s

18. Belyakov G «About Flying Humanoids», NLO, 2010,16

19. Leuter K, »The Devil Hunters» – OFFICIAL RESEARCHERS OF THE JERSEY – Continuing Edition – http://www.njdevilhunters.com

20. «Reptilian Humanoid Startles Oklahoma Dispatcher», CRYPTOZOOLOGY NEWS, 2018 – http://cryptozoologynews.com

21. «The Chupacabra got to Kharkov?» PARANORMAL - NEWS.RU 26.05.2010

22. http://www.ruarchive.com/archives/10075

23. «Aggressive Indian Invisibles» – PARANORMAL NEWS 12.09. 2015

24. «Kuban News», 05/22/2002, cit. by Kolchin G «UFOs and aliens: invasion of the earth», 2006

25. «News of the Ukraine» TC «1+1» 20.07. 2011

26. Mishin D, «Chupakabra» - http://world.lib.ru/m/mishin_denis_andreewich/chupakabra-1.shtml

27. Grebennikov VS, Zolotarev VF «Theory of field radiation of multicavity structures» – Abstracts of the report at the interdisciplinary scientific and technical school-seminar of the Tomsk Polytechnic Institute, 18-24.04.1988

28. Grebennikov VS «My World», 1997 (in Russian)

29. Downes J, Davies G «Owlman and Others» , 1997

30. «The Chupacabra returns exactly one year later», «UKRAINE ABNORMAL» 06/29/2012

31. McManus DM «Folklore of the counties of England» cit. by Nepomniachtchi N «100 Great Mysteries of Nature», 2003

32. Sergushev M, FACTI (the Ukraine) 18.07.2009 (in Russian)

33. TV plot «Residents of Buryatia again report attacks by the mysterious «chupacabra», ARIGUS-TV NEWS 01/22/2014

34. «Mysterious creature sighted at Khurai, Manipur.Cat hatlmme hyye loin khottlaga», Yutube Channal THE INVINCIBLE AUTOCRAT 29.11.2018 https://www.youtube.com/watch?v=1a13oG496Hk&list=FLBwQ9NQbmq0Ke4KkoJoxs_A

35. ARGUMENTI I FACTI (Altai) 18.07.2011 (in Russian)

36. Efimov V «Ants Atta: Teleportation» – https://www.13min.ru/drugoe/muravi-atta-teleportatsiya/;

37. Butusov K «Ants atta and teleportation» – PARANORMAL NEWS 09/04/2013

38. McDonald C, Kaplan J »Prague in the Shadow of the Swastika: a History of the German Occupation 1939-1945», 1995

39. Tsareva A «These mysterious animals» www.rulit.me/books/eti-zagadochnye-zhivotnye-read-181395-2.html

40. «Firs Sighting of the Mothman»- West Virginia Department of Commerce # 1215

41. Lyashenko V «Chupacabra in the Ukraine» - https://boristen70.livejournal.com/18383.html

42. «Chupakabra continues atrocities in the Vinnitsa region», UKRAINE ABNORMAL 17.04.2012

43. FACTI (Ukraine) 26.03.2010

44. Sergushev M, FACTI (Vinnitsa-Kyiv) 05/14/2010,

45. Bykov A»Chupakabra: in the Kuzbass village of Bekovo the horror settled?» 02/07/2019 – http://avoka.do/posts/chupakabra-v-kuzbasskom-sele-bekovo-obosnovalsya-uzhas

46. «BOGDANCLUB», «Chupacabra» – http://bogdanclub.info

47. «Chupakabra in the Chernihiv region scared a woman in the cemetery» – WEBSITE OF THE CITY OF CHERNIGOV – https://www.0462.ua/news/456831/cupakabra-v-cernigovskoj-oblasti-napugala-zensinu-na-kladbise

48. Rudakova O «KYSHTYMSKI RABOCHI», 09/25/1996

49. Gershtein M, Deruzhinsky V «Undead or Mysterious Creatures», 2015

50. Finucane RC «Appearances of the Dead: A Cultural History of Ghosts», London, 1982

51. Kolchin G «UFO, facts and documents», 1991

52. Turner K «Into the fringe», 1992
53. Turner K, Rice T «Masquerade of Angels», 1994
54. Turner K «Taken: Inside the Alien-Human Agenda», 1994
55. Zakharia Sitchin, Giorgio A. Tsoukalos, Erich von Daniken, Andrey Sklyarov, Alexander Koltypin wrote about paleocontacts.
56. Pereira JU «Les «Extra-terrestres», GEPA, 1969, 2, P. 3-72
57. Randles J, «UFO study. A Handbook for Enthusiasts», 2007
58. «Testimony of Ezekiel, son of Buziah, a priest, in the land of the Chaldees, by the river Chebar.» Synoid translation of the Bible
59. Kramer SN «History begins in Sumer», 1965
60. «Umbra: Demonology as a semiotic system». ALMANAC, vol. 2, 2013
61. «Vampire in Europe. Montague Summers», Aquarian Press, 1980
62. Makhov AE «Hostis Antiquus: Categories and Images of Medieval Christian Demonology», M, 2006
63. Finucane RC «Appearances of the Dead: A Cultural History of Ghosts», London, 1982
64. Shawn Cipa «Carving gargoyles grotesques, and other creatures of myth: history, lore, and 12 artistic patterns», 2009
65. Tylor EB «Primitive Culture», 1989 (in Russian)
66. Belova O, Pertukhin V «Jewish myth in Slavic culture», LITRES, 2018
67. Gumilyov LN «Ancient Russia and the Great Steppe», 2012
68. Klimov VV – «Notes to the legends about the Chud» // Questions of linguistic local history of the Kama region. Issue. 1. Perm, 1974. P. 121-123 (in Russian)
69. Dark Secrets inside Bohemian Grove by Alex Jones – https://youtu.be/wtSVBTne-KY
70. Bord J, Bord C «Alien Animals, PANTHER GRANADA, 1985

71. Argentine website Vision Ovni, founded in 1991, is an Argentine national UFO research organization. The news has been archived as number 1391 – www.visionovni.com

72. Redfern N «The Monster Book: Creatures, Beasts and Fiends of Nature», VISIBLE INK PRESS, 2016

73. «People-birds in the history of mankind», «PARANORMAL NEWS» 03/06/2014

74. . Pavlovich IL, Ratnik OV «Secrets and legends of the Volga dungeons», 2003 (in Russian).

75. Turner K «Encounter Phenomena Defy «Set Pattern», UFO, 1993, 8,1

76. Maltsev S. «Invisible battle. Secret history of civilization», 2003

77. Howe LM «Glimpses of Other Realities», V2, 1998

78. Leir R. MUFON UFO JOUJNAL, Mars 1998

79. Valle J «Passport to Magonia : on UFOs, folklore, and parallel world», 1993

80. Hall Richard, Keyton D., Gough D. «UFO and NATO. «The human mutilation is cover up» – Richplanet TV Production, 2014; Part 1 and 2 – richplanet.net

82. Romme M, Escher S «Accepting Voices», 1998

83. Fomenko VN «The Earth as we do not know it», 2001

84. «These children hear creepy voices: how to help them?» BBC News Russian Service – 8 June 2018

85. Charles Fernyhough, Alien voices and inner dialogue: towards a developmental account of auditory verbal hallucinations,New Ideas in Psychology, Volume 22, Issue 1, 2004, Pages 49-68

86. Titus Livy «War with Hannibal», 1993

87. Togoeva OI «Shepherdess who became an Amazon. The image of Joan of Arc in the French biographies of famous people of the 16th-17th centuries» // ADAM AND EVE. ALMANAC OF GENDER HISTORY, 2012, Issue 20, P 11-36

88. Blavatsky EP «Isis Unveiled», vol 1, 2, 2022

89. Goodrick-Clark N «The Occult Roots of Nazism», 2004

90. Balagushkin EG, Nontraditional Religions of Modern Russia, 1999

91. ALTERNATE PERCEPTION № 52, 2001

92. ALTERNATE PERCEPTION № 53, 2002

93. Corrado Malanga «Gli UFO nella mente», 2016

94. Anfalov AA «Aliens in the Crimea», 2018 (in Russian)

95. Giles Whittell, «UFO cult aims to clone humans for childless and gay couples», 2000

96. Azhazha V «Another Life», 1981

97. Approximately the same ideas in 1989 were voiced by Yuri Krivonogov, the future leader of the White Brotherhood, at lectures organized at Kiev Institute of Civil Aviation Engineers. From the personal observations of this book author, who visited this lecture hall in the autumn of 1989, the lecture hall scool was organized by Kyiv ufologists. For a long time, the author was perplexed about this experiment, allegedly carried out by the KGB. However, later it turned out that foreign (or «alien») intelligence services were involved in it.

99. Maccabee B «The UFO/FBI Connection», Llewellyn Books, 2000

100. «Briefing Document: Operation MAJESTIC 12» 18 Nov 1952

101. CIENTIFIC STUDY OF UNIDENTIFIED FLYING OBJECTS, Conducted by the University of Colorado, Under contract No. 44620-67-C-0035 With the United States Air Force, Dr. Edward U. Condon, Scientific Director, 1968 (Internet Edition Prepared by National Capital Area Skeptics (NCAS) District of Columbia – Maryland – Virginia (USA), Release 1 – January 1999

102. Corso Ph. J, Birnes W. «The Day After Roswell», 1998

103. According to the memoirs of Professor V. Burdakov, back in 1947, leader of the USSR I.V. Stalin consulted on this issue with S.P. Korolev – Shelepov V. «UFO: Operation Cover-up» // «TOP SECRET» 10-6-97 (in Russian) . .. According to some information, approximately as early as 1948 in the USSR there was a project to create weapons that would make it possible to resist aggression or hostile actions from space. Like nuclear projects, this was also led by Lavrenty Beria, but the level of secrecy was an order of magnitude higher. Link [14] from part 2 to the creation of weapons against «aliens»

104. Milton William Cooper, «Behold a Pale Horse», 1991

105. Back in the years of World War II, American aircraft pilots noticed strange objects that often accompanied them during

flights to bombard targets in Germany and Japan. On August 6, 1945, the Americans dropped an atomic bomb on Hiroshima. A little later, many pictures were taken of what was left of Hiroshima after the bombing. Upon closer examination, some of them found an unidentified object, «hanging» over the ruins for a week. On June 25, 1950, the Americans began fighting in Korea. And already on June 26, UFOs could often be seen in the sky of Korea, which accompanied the planes and hovered over the combat areas. The same pattern was observed in Vietnam. So, in June 1967, American observers spotted a disk-shaped UFO, which had a large size. Two squadrons of fighters were raised to intercept the object, but it was enveloped in an incomprehensible cloud and went up at high speed, disappearing from the visibility zone. Repeatedly unidentified objects were seen over Hanoi. They often appeared over American aircraft carriers. Having taken out about a hundred V-2 missiles from Germany, the Americans began test launches. When one of the rockets was launched in 1948, a disk-shaped UFO suddenly appeared and began to make circles around the flying rocket, and then, having developed a speed of about 9 thousand kilometers per hour, overtaking the rocket, went into outer space. Subsequently, similar cases were recorded in April and July 1949, as well as in April and July 1950. When a UFO was sighted during a Polaris rocket launch from Cape Canaveral in 1962, the range's photographers were instructed to track and photograph the appearance of unidentified objects during the rocket launch. An interesting incident occurred in September 1964 during the launch of the Atlas rocket from the Vandenberg base. After the launch, high-speed shooting of the rocket was carried out using a telescope. Looking at the film after the end of filming, the specialists saw that a disk-shaped UFO approached the missile warhead flying at an altitude of about 110 km and circled it twice, emitting four bright flashes, after which it left. Literally a few seconds after that, the warhead went out of control and, before reaching the target, fell into the ocean. On March 17, 1950, a total of over 500 unidentified objects flew over US nuclear weapons factories located in New Mexico. Cases of hovering of unidentified objects have also been documented over other American plants related to the production of nuclear weapons: in Oak Ridge, Hanford, Los Alamos, Las Vegas. In 1965, a UFO was seen hovering over the British Nuclear Research Establishment. Unidentified objects and Russian strategic objects did not bypass their attention. Their appearances over Baikonur and Kapustin Yar occurred quite often. In the fifties of the last century, UFOs regularly appeared over the nuclear test site near Semipalatinsk. Moreover, according to eyewitnesses, they appeared over the site of a future explosion just a few seconds before the laying of the charge in the mine. Military depots also attract their attention. In 1978, a luminous spherical object appeared above a military warehouse located in the Pskov region. Hovering in turn over each vault, the ball directed a beam of light at it, and then, with a jerk, moved to the next one, right in front of the entire guard. At the same time, the ball appeared the next day, repeating its actions. An anti-aircraft machine-gun installation was urgently placed near the storage facilities, but the UFO did not fly again. – From an article by Viktor Romanchenko, «SENSATION – Russian military declassify their UFO archive», 2000

106. Von Wettenburg B «Gehenime Ufo-Sache. Schach der Erge», 1996

107. Pacaccini V, Ports M «Incidente em Varginha: criaturas do espaço no Sul de Minas», 1996

108. According to A. Anfalov – see Appendix 4 in the book Agon E., Anfalov A. «Heavenly angels or dogs of hell?» , 2022 (in Russian)

109. By Branton «The Dulce Wars: Underground Alien Bases and the Battle for Planet Earth», 2011, 135 pages

110. AFOSI – Office of Criminal Investigation of the United States Air Force – the investigative and counterintelligence agency of the United States Air Force and intelligence

111. It was about experiments with space-time carried out by American specialists at test sites in Siberia. kxk.ru/energy/v16_608582__1294249525.php

APPENDIX 1

REPORTE-96:
«ANATOMY AND PHYSIOLOGY OF ALIEN»

www.x-libri.ru/elib/innet237/index.htm

SECRET. SPREAD -
SIGNIFICANT SIMILARITY WITH THE AVAILABLE-
IN 1965 AS A RESULT
COLLED AS DANGEROUS
CONTAINING GENERAL REVIEW -
BASED ON THE SYMBIOSIS OF ONE
THIS SECTION IS CONSIDERED -
CELLULAR BILATERAL-SYMMETRIC -
COLONIAL WORMLIKE --
BILATERAL SYMMETRY --
TENTACLE VESICLE. DUE TO CLIMATE CHANGE
EXOSKELETAL STRUCTURES, INCLUDING LIMB -
LOCOMOTIONS, AND THE PROCESS OF ENCEPHALIZATION BEGAN -
ORGANS NOT INVOLVED IN THE FORMATION -
TURNED INTO RESPIRATORY AND VOCALIZATION ORGANS -
WEAK DEVELOPMENT OF DISTANT FORMS OF RECEPTION
INFORMATION ABOUT EXTERNAL -
MAINLY THROUGH CHEMORECEPTORS AND CEREBRAL-
IN PARTICULAR, THE TIME RELATIONSHIP BETWEEN THE APPEARANCE
OF PRENATAL SURGERY AND ALLOCID-DEPENDENT TRANSDUCT.

THE GENERAL ANATOMY OF ALIENS WILL BE CONSIDERED BY EXAMPLE

....

VERY LARGE SIZE. FOR EXAMPLE, AN ADULT SMALL WORKER REACHES 2.5 METERS IN LENGTH (WITH ELONGATED LOWER FIXTURE OF THREE WELL-PRONOUNCED DEPARTMENTS.

THE TOP DEPARTMENT (LOFOFOR) IS AN NON-SEGMENTED CHITINIZED TUBE, TOPPED WITH SIX PERISTIVE CHEMORECEPTOR APPENDAGES, ANTENNULES,

INCLUDING SO-CALLED CUPULA PHEROMONE RECEPTORS IN THE UPPER LOPHOPHORE

…………..

ALSO ARE TWO SIMPLE EYES, CONSISTING OF A COVER LENS, INNER VITRIC MASS AND A SENSOR PLATE. UPPER LIMBS ARE LOCATED ON THE SIDES OF THE LOPOPHORUS, EACH OF WHICH HAS TWO SECTIONS: A MUSCULAR TUBE AND A TENTAL COLORL OF 10 – 15 TENTACLES.

LOFOFOR UPPER LIMB ENDOSKELETON IS ABSENT. THE EXOSKELETON CONSISTS OF ELLIPTICAL PARTS CONNECTED WITH HUNDREDS OF TILE-SHAPED PLATES OF DIFFERENT SIZES AND SHAPES, UPPER AND LOWER. THEY COVER THE MUSCLE TUBE AND EACH OF THE TENTACLES. TILE-SHAPED PLATES ARE CONNECTED BY SHELLED LINKS FROM THINER CHITIN. THANKS TO SUCH A DEVICE EACH POINT OF THE UPPER LIMB HAS SIX DEGREES OF FREEDOM. CLOSING TEETH LOCATED ON THE DISTAL SURFACES OF THE TILE-SHAPED PLATES FIX THE LIMB IN A CERTAIN POSITION. THEIR CLOSURE IS CARRIED OUT BY SPECIAL MUSCLES (MM. FIXATORES LIMBI SUPERI

THE PRE-SKELETAL BASIS OF THE LOPHOPHORE IS THE LANCE-SHAPED PROCESS OF THE CHORDAL RING (SEE BELOW), PASSING THROUGH THE CONNECTING HOLE INTO THE BRANCHIOPHORE. ANTENNAL ARTERY, THE BRANCHES OF WHICH FEED THE UPPER LIMBS, ANTENNAS AND SIMPLE EYES, ANTENNAL VEIN, RESPONSIBLE FOR THE VENOUS OUTFLOW FROM THESE AREAS, A POWERFUL SPELL-LIKE NERVE PLEXUS WITH A SPIRAL AROUND A SPEAR-SHAPED TRANSFORMATION OF NINETEEN-TWENTY TWO NERVE GANGS I ORDER (SPIRAL GANGWAYS), RESPONSIBLE FOR THE SENSITIVE AND MOTOR INNERVATION OF THE UPPER LIMB, LUNGS AND PSEUDOSYRINX (SEE BELOW), AS WELL AS THE ANTENNULAR AND OPECULAR NERVES.

THE MIDDLE DEPARTMENT (BRANCHIOPHORE) HAS THE FORM OF FOUR CHITIN BUBBLES (BAG) CONNECTED AT THE BASIS. AT THE TOP OF EACH THEC THERE ARE HOLES COVERED WITH FILTRATION FOLDS. SMALLER, UPPER BUBBLES – SYRINGOTEKS – CONTAIN A PAIR ORGAN OF VOCALIZATION – PSEUDOSYRINX. LARGE, LOWER BUBBLES – BRANCHIOTHECAS CONTAIN PAIR «LUNGS» OF ALIENS.

ON THE FRONT SURFACE OF THE BRANCHIOPHOR, AT THE PLACE OF CONFERENCE OF THE BASES OF THE THECA (DECUSSATIO THECARUM), THERE IS A MOUTH HOLE, SURROUNDED BY A GRASPING-GASPANT ORGAN. THE LAST (IN THE SMALL WORKER) CONSISTS OF FOUR CHITIN HOOKS ABOUT 25 CM LONG, CONNECTED BY A COMMON FIBRO-MUSCULAR MEMBRANE. FROM THE MOUTH, THE PHARYNAL STARTS, GOING DOWN AND BACK AND ENDING IN THE AREA OF THE UPPER SFINCTER OF THE STOMACH. BEHIND THE PHARYNGEAL IS THE BASIS

OF THE SPONE-SHAPED PROCESS, CONTINUED UP INTO THE LOPOPHORUS, AND THE VESSELS AND NERVES ASSOCIATED WITH IT (SEE ABOVE).

THE LOWER SECTION (VISCEROTHECA) IS AN UNSEGMENTED ELLIPSOID FROM DENSE CHITIN. IN THE AREA OF THE UPPER POLE OF THE VISCEROTHECA IS THE CHORDAL RING – THE BASIS OF THE ALIENS' ENDOSKELETOUM. THIS IS A DOUBLET-SHAPED FORMATION CONSTRUCTED FROM OSSEOID FABRIC. ON THE LOWER SURFACE OF THE CHORDAL RING A GENERAL MEMBRANE IS ATTACHED – A FIBROUS FILM LINED WITH MESOTHELIA. THE GENERAL MEMBRANE CONSISTS OF 5 DEPARTMENTS: 1) MEMBRANE OF THE GASTRIC; 2) MEMBRANE OF THE HEART RINGS; 3) MEMBRANE OF THE GREEN GLAND; 4) MEMBRANE OF THE SMALL INTESTINE; 5) MEMBRANE OF THE COLON.

THE MAIN BLOOD VESSELS PARACHORDAL (LOWER) AND BRANCHIAL (UPPER) RINGS ARE INSIDE IN THE THICKNESS OF THE COMMON MEMBRANE. FROM THE INTERNAL SURFACE OF THE CHORDAL RING THE HIALINE PLATES DEPART, FORMING THE CAPSULE OF THE BRAIN. ORGANS SUSPENDED ON THE MEMBRANE FILL ALL THE INTERNAL SPACE OF THE VISCEROTHECA. IN THE AREA OF ITS LOWER POLE IS THE ANUS, WHERE THE COLON AND THE DUCT OF THE GREEN GLAND OPEN

THE UPPER LIMB, LOCATED ON THE SIDES OF THE VISCEROTHECA, DO NOT HAVE ENDOSKELETOUM.

THE MYOEPITHELIAL BASIS OF THE LOWER LIMB CONSISTS OF TWO DIVISIONS – MUSCULAR TUBE AND TERMINAL EXPANSION. THE EXOSKELETON OF THE LOWER LIMB, UNLIKE THE UPPER LIMB, IS REPRESENTED BY AN ELLIPTICAL ARTICULATE ELEMENT, TWO TUBULAR FORMATIONS A THIGH AND A SHIN CONNECTED BY TWO TRANSITION PLATES. THE DENSE CUTICLE OF THE TERMINAL EXPANSION IS PIERCED BY STRIPS ENDING IN POWERFUL SUPPORT SPURS.

THE ALIEN ENDOSKELETON CONSISTS OF TWO BASIC FORMATIONS OF THE TRIANGULAR CHORDAL RING AND THE LAST LANCER PROCESS EXTENDING VERTICALLY FROM THE UPPER SURFACE.

THE INTERNAL SURFACE OF THE CHORDAL RING CONTINUES INTO THREE THIN TRIANGULAR HIALIN PLATES. THE BRAIN CAPSULE IS A DUPLICATE OF HIALIN PLATES. AT THE PLACE OF DISCHARGE OF THE SPONE-SHAPED PROCESS FROM THE CHORDAL RING THEY ARE SEPARATED BY A TRIANGULAR SLIT THROUGH WHICH THE ANTENNULAR, OBLIGAL, COMMON VISCERAL NERVES PASS, AS WELL AS THE ROOTS OF THE SPONE-SHAPED AND PEDAL PLEXES.

……… CONSTRUCTED FROM OSSEOID FABRIC – MICROSCOPIC CRYSTALS OF ARAGONITE AND IRON CARBONATE, SOLDERED BY STRUCTURELESS SULFATED

POLYSACCHARIDES AND FIBRILLAR KERATIN-LIKE PROTEIN. CELLULAR ELEMENTS – CHORDOBLASTS – ARE ONLY IN THE THIN (100-200 μM) SURFACE LAYER. MINERAL SPICULE IN HYALINE PLATES

EXTERNAL PROTECTIVE ORGANS OF ALIEN ARE REPRESENTED BY MYOEPITHELIAL TISSUE OF THE I KIND. EXOSKELETONAL STRUCTURES ARE A PRODUCT OF THE SECRETION OF THE MYOEPITHELY. ALL OF THEM ARE PERFECTED WITH THIN CHANNELS CONTAINING CYTOPLASMATIC PROCESSES OF MYOEPITHELIOCYTES. BASAL CONTRACTILIVE FIBERS OF MYOEPITHELIOCYTES FUSE TOGETHER, FORMING THE MUSCLE BASIS OF THE LIMB AND THE GENERAL SUBCUTICULAR MUSCULAR LAYER, AVAILABLE IN ALL SECTIONS OF THE BODY, BUT ESPECIALLY DEVELOPED IN THE LOWER PART OF THE LOPHORUS. INSIDE THE SUBCUTICULAR MUSCLE LAYER IS LINED WITH MESOTHELIA. IN THE VISCEROTHECA, BETWEEN THE SUBCUTICULAR LAYER AND THE MESOTHELIUM, THERE IS A LOBULATE ACCUMULATION OF FAT CELLS (FAT BODY).

RESPIRATORY AND VOCALIZATION ORGANS OF ALIENS ARE REPRESENTED BY A PAIR PSEUDOSYRINX (VOCALIZATION ORGAN), LOCATED IN THE TWO UPPER BAGS OF THE BRANCHIOPHER, AND PAIR «LUNG» IN THE TWO LOWER BAGS.

BOTH PSEUDOSYRINKS AND «LUNGS» ARE MYOEPITHELIAL ORGANS FORMED BY A NETWORK OF TRABECULS. INSIDE EACH TRABECULA ARE VESSELS AND CONTRACTIBLE FIBERS. OUTSIDE TRABECULA COVERED WITH MYOEPITHELIA OF THE IV KIND. TRABECULS OF THE «LUNGS» HAVE A DIAMETER OF ABOUT 50 MKM, AND THE THICKNESS OF THE AIRHEMALE BARRIER IS 15 MKM. PSEUDOSYRINX TRABECULS ARE SIGNIFICANTLY THICKER (FROM 500 μM TO 7 MM) AND HAVE A PROGRESSIVE FRAME OF ELASTIC FIBERS. IN ADDITION, PART OF THEM IS SUPPLIED WITH CHITIN COMBS AND HOOKS.

EACH MYOEPITHELIAL ORGAN HAS A ROOT AT THE BASIS OF THE CORRESPONDING BAG. THE MAIN VESSELS PASS IN THE REGION OF THE ROOT – BRANCHES OF THE BRANCHIAL RING – AND A DRAINAGE DUCT BEGINS, OPENING INTO THE THROAT AT THE BASIS OF THE GRASPITABLE ORGANS. THE DRAINAGE DUCT IS A TUBE ABOUT 30 MM LONG, FORMED BY TYPE IV MYOEPITHELIA.

BLOOD CIRCULATION ORGANS.

THE CIRCULATION SYSTEM OF ALIENS IS A CLOSED TYPE. ITS THE MAIN ELEMENTS ARE TWO LARGE RING VESSELS – PARACHORETAL AND BRANCHIAL RINGS – LAID IN THE THICKNESS OF A COMMON MESENTERY, TWO CONNECTING DUCTS AND EMBROILING MUSCLE PILLOWS INCLINED IN THEIR WALLS, PARACHORETORY, THE RING CONSISTS OF ARTERIAL AND VENOUS SEGMENTS. THE DIAMETER OF THE VENOUS SEGMENT IS LESS THAN THE ARTERIAL. FROM

THE ARTERIAL SEGMENT THE ARTERIES SUPPLYING THE PHARYNX, THE STOMACH, THE INTESTINES, THE GREEN GLAND, IN UPPER CASTS – THE GONADS. INTO THE VENOUS SEGMENT THE RELATED VEINS FLOW.

THE UPPER, BRANCHIAL, RING CONSISTS OF TWO SEGMENTS: BRANCHIAL (FROM THE PLACE OF CONCENTRATION OF THE VENOUS CONNECTING DUCT TO THE PLACE OF CONCEPTION OF THE PULMONARY VEINS) AND ANTENNALE (FROM THE PLACE OF CONCEPTION OF THE PULMONARY VEINS TO THE PLACE OF DISCHARGE OF THE ARTERIAL CONNECTING DUCT). THE CEREBRAL, SYRINGOID AND ANTENNALE VEINS FLOW INTO THE BRANCHIAL SEGMENT, THAT IS FLOWING WITH NON-OXYGENATED BLOOD, THEN TWO PULMONARY ARTERIES PARTIATE FROM IT (ONE TO EACH LUNG). TWO PULMONARY VEINS, CARRYING OXYGENATED BLOOD FROM THE LUNGS, FLOW INTO THE ANTENNA SEGMENT. THE CEREBRAL ARTERY (TO THE BRAIN), THE SYRINGOID ARTERY (TO THE PSEUDOSYRHYNXUS) AND THE ANTENNA ARTERY (TO THE LOPOPHORUS ORGANS) GO FROM IT. THE DIAMETER OF THE BRANCHIAL RING NARKS AT THE PLACES OF THE VENTURES AND EXPANDS AT THE PLACES OF THE ARTERIES.

THE VENOUS CONNECTING DUCT CONNECTS THE VENOUS SEGMENT OF THE PARACHORDAL RING AND THE BRANCHIAL SEGMENT OF THE BRANCHIAL RING.

THE ARTERIAL CONNECTING DUCT CONNECTS THE ANTENNA SEGMENT OF THE BRANCHIAL RING AND THE ARTERIAL SEGMENT OF THE PARACHORDAL RING.

EACH DUCT IS SURROUNDED WITH A «CLUTCH» FROM CIRCULARLY ARRANGED SMOOTH MYOCYTES. THE CONNECTING DUCTS AND THEIR MUSCLE «COUPLINGS» TOGETHER ARE IN THE AREAS OF THE PARACHORDAL AND BRANCHIAL RINGS LOCATED BETWEEN THE ORIGINS OF THE DUCTS (INTERDUCTORAL, OR THIRD, SEGMENT) IN THE WALLS OF THE VESSELS CIRCULAR PLATES ARE ALSO PLACED. THEIR FUNCTION IS REGULATION OF THE DEGREE OF MIXING OF ARTERIAL AND VENOUS BLOOD.

DIGESTIVE ORGANS OF ALIENS ARE REPRESENTED BY THE PHARYNSE, EGROUS STOMACH, SMALL AND LARGE INTESTINES. ALL THE SPECIFIED FORMATIONS ARE FORMED BY MYOEPITHELIAL TISSUE II AND III TYPE. THE MUSCLE COMPONENT OF THE MYOEPITHELIUM FORMES A FUNCTIONAL SYNCITIUM WITH A CROSSBOWED CONTRACTILITY FIBERS RESPONSIBLE FOR PERISTALTIS.

THE PHARYNGEAL IS FORMED BY MYOEPITHELIAL TISSUE OF THE II TYPE AND IS A TUBULAR FORMATION ABOUT 25 CM LONG, OPENING INTO THE MUSCLE STOMACH IN THE AREA OF ITS UPPER SPHINCTER. IN THE UPPER DEPARTMENT OF

THE THROAT, IMMEDIATELY BEHIND THE GRASPING AND EATING ORGANS, THE LUNG AND PSEUDOSYRINX DRAINAGE DUCTS OPEN, TWO ON EACH SIDE.

THE STOMACH IS FORMED BY MYOEPITHELIAL TISSUE OF THE II TYPE AND IS A SAC-LIKE FORMATION WITH A CAPACITY OF UP TO 20 LITERS. IN THE AREA OF THE SPINCH OF THE PHARYNGEAL AND THE OUTPUT OF THE SMALL INTESTINE THE MUSCULAR COMPONENT OF THE MYOEPITHELIA FORMS TWO CIRCULAR THICKENINGS – THE UPPER AND LOWER SPHINCTERS.

THE SMALL INTESTINE IS FORMED BY MYOEPITHELIAL TISSUE OF THE III TYPE AND IS A TUBE ABOUT 1 M LONG. OUTSIDE OF THE MUSCLE SYNCITIUM ARE NUMEROUS ENCOUNTERED IN THE CONNECTIVE TISSUE CAPSULE GLANDS PRODUCING A HYDROLASE MIXTURE. THEIR EXECUTIVE DUCTS OPEN ON THE INTERNAL SURFACE OF THE INTESTINE.

THE COLON IS FORMED BY MYOEPITHELIAL TISSUE OF THE II TYPE AND REPRESENTS A TUBE WITH A RELATIVELY SMOOTH INNER SURFACE ABOUT 10 M LONG. AROUND THE LAST MUSCLE COMPONENT OF THE MYOEPITHELIA FORMS A CIRCULAR THICKENING – THE SPHINCTER. BELOW THE LOWER EDGE OF THE SPHINCTER, THE GREEN GLAND DUCT IS OPENED BY A SEPARATE HOLE.

THE ORGAN OF DETOXICATION AND MAINTENANCE OF SALT HOMEOSTASIS – THE GREEN GLAND – IS A SHAPED GREEN LOBE FORMATION WEIGHT ABOUT 3 KG, FORMED BY MYOEPITHELIAL TISSUE OF THE IV KIND. THE SECRET OF THE GREEN IRON IS A HIGH MOLECULAR POLYSACCHARIDE HAVING THE PROPERTIES OF A UNIVERSAL SORBENT FOR CATIONS, ANIONS AND POLAR ORGANIC SUBSTANCES WITH ONE OR SEVERAL AROMATIC NUCLEI. WITH THIS SECRET, BOTH EXCESS SALT AND NON-METABOLIZABLE ORGANIC TOXINS ARE RECOVERED.

THE NERVOUS SYSTEM OF ALIEN CONSISTS OF A CENTRAL SECTION (BRAIN) AND A PERIPHERAL SECTION (ANTENNULAR, OBLIGAL, COMMON VISCERAL NERVE, LANCER, SPIRAL PLEXUS, ETC.).

THE BRAIN OF ALIENS, AS STATED ABOVE, IS IN A CAPSULE FORMED BY THE CLEAVAGE OF HIALIN PLATES. THE TRIANGULAR SLOT OF THE CAPSULE IS TIGHTENED WITH A CONNECTIVE TISSUE MEMBRANE. THE CAPSULE IS FILLED WITH A LIQUID, DENSITY IS CLOSE TO THE BRAIN SUBSTANCE.

MACROSCOPICALLY THE ALIEN BRAIN HAS FIVE WELL-DESIGNED SECTIONS.

1.. THE VISCERAL BRAIN – A GLOBULAR FORMATION WITH A SMOOTH SURFACE, LOCATED IN THE BACK-BOTTOM PART OF THE BRAIN CAPSULE. FROM THE UPPER SURFACE OF THE VISCERAL BRAIN THE UNPARATED COMMON VISCERAL NERVE IS DEPARATED. THE LATERAL SURFACES ARE CONNECTED TO THE STEM OF THE SOMATIC BRAIN VIA TWO CONNECTIVES. ON THE CUT THE

VISCERAL BRAIN CONSISTS OF: 1) OUTER LAYER OF GRAY SUBSTANCE ABOUT 1.5 MM THICK (FORMED BY NEUROCYTE BODIES); 2) A LAYER OF WHITE MATTER, ABOUT 5 CM THICK, FORMED BY NEUROCYTE PROCESSES; 3) THE CENTRAL GRAY NUCLEUS, FORMED BY THE BODIES OF NEUROCYTES AND CONSISTING OF SEVERAL DIFFICULTLY DISCOVERED CELL GROUPS.

THE FUNCTION OF THE VISCERAL BRAIN IS THE SENSITIVE AND MOTOR INNERVATION OF INTERNAL ORGANS, AS WELL AS THE CONTROL OF METABOLISM, ENDOCRINE FUNCTIONS AND SECRETION OF PHEROMONES, CARRIED OUT MAINLY, NEUROSECRETORY ELEMENTS OF THE CENTRAL GRAY NUCLEUS AND INNERVATING EXO-AND ENDOCRINE GLANDS FIBERS OF A COMMON VISCERAL NERVE.

2. SOMATIC BRAIN _ – FORMATION OF IRREGULAR CYLINDRICAL FORM WITH THREE THICKENINGS, SYRINGOBRONCHEAL, TENTAL AND PEDAL. THE SOMATIC BRAIN IS IN THE BACK OF THE BRAIN CAPSULE.... ON THE LATERAL SURFACES THERE ARE THICKNESSES LOCATED ROOT AND TRUNKS OF NERVES CORRESPONDING TO BODY SEGMENTS. EACH OF THEM IS FORMED BY THE MERGING OF THE UPPER (SENSITIVE) AND LOWER (MOTOR) ROOTS. FROM THE TENTACLE THICKENING, TWO FROM THE SYRINGOBRANCHIAL – FIVE AND FROM THE PEDAL – THREE SEGMENTAL NERVE ON EACH SIDE. SEGMENTAL NERVES EXIT THE BRAIN CAPSULE THROUGH HOLES ON ITS UNDERSIDE. TWO SEGMENTAL TENTAL AND FIVE SEGMENTAL SYRINGOBRANCHIAL NERVES GO BACK FROM BRAIN CAPSULE TO THE PARACHORDAL RING, PASSING THROUGH THE HOLE IN THE COMMON MESSENUM, AND REACH THE BASIS OF THE SPELLITED PROCESS. THERE THEY FORM A SPONE-SHAPED INTEXTURE. THREE PEDAL NERVES GO TO THE LOWER INTERNAL SURFACE OF THE VISCEROTHECA, WHERE THE PEDAL PLEXUS IS FORMED.

ON THE CUT THE SOMATIC BRAIN CONSISTS OF A PERIPHERAL LAYER OF GRAY SUBSTANCE ABOUT 2 MM THICK AND A CORE FORMED BY WHITE SUBSTANCE. IN THE AREA OF THICKNESS, THE THICKNESS OF THE PERIPHERAL LAYER REACHES 4 MM, THE TRANSITION GROOVE DIVIDES IT INTO TWO CLEARLY DIFFERENT PLATES – A LARGE (SENSORY AND ASSOCIATIVE) AND A SMALLER LOWER (MOTOR).

3. SENSORY BRAIN IS A PAIR FORMATION LOCATED IN THE FRONT UPPER PART OF THE BRAIN CAPSULE. CONSISTS OF TWO LOBS (VISUAL AND ANTENNULAR), SEPARATED BY THE BORDER FUROUS. FROM THE UPPER SURFACE OF THE SHARE THE EYE AND THREE ANTENNULAR NERVES. HALVES OF THE SENSORY BRAIN ARE CONNECTED BY A ROLLER WITH A WIDE FLAT BUNCH OF NERVE FIBERS. FROM THE LOWER SURFACE OF THE ROLLER ARE TWO FRONT CONNECTIVES – A LONG THIN BUNCH OF FIBERS WHICH GO DOWN TO THE

TERMINAL BRAIN, TWO MIDDLE CONNECTIVES GOING TO THE MIDDLE BRAIN, AND TWO REAR CONNECTIVES GOING BACK TO THE MEDIA BACK TO THE MEDIA.

4. THE OLIENCEPHALON BRAIN IS A CONTINUATION OF THE STEM OF THE SOMATIC BRAIN AND OCCUPATES THE CENTRAL PART OF THE BRAIN CAPSULE. THE BASIS OF THE INTERNET BRAIN IS MADE UP BY THE CENTRAL FIBER BUNCH – A POWERFUL LAYER OF NERVE FIBERS BETWEEN WHICH ARE PLENTY (UP TO FIFTY) CELL CLUSTER – INTERNECULAR NUCLEI FORMING THE SENSOR PART OF THE CEREBRAL MAGNETOSENSOR. OUTSIDE, THE CENTRAL FIBER BUNCH IS COVERED WITH A LAYER OF GRAY SUBSTANCE UP TO 3 MM THICK – A CONTINUED GRAY SUBSTANCE OF THE SOMATIC BRAIN STEM. CELLS OF THIS LAYER ARE RESPONSIBLE FOR THE ORGANIZATION OF COMPLEX MOTOR ACTS. TWO CONNECTIVES FROM THE SENSORY BRAIN ARE SUITABLE TO THE LATERAL SURFACE OF THE OLIENCEPHALON BRAIN. ON THE BACK SURFACE OF THE INTERBRAIN THERE IS A CAVITY NAMED MAGNETOSENSOR BULLA. IT IS FILLED WITH A COLORLESS TRANSPARENT LIQUID – MAGNETOLYMPH. INSIDE THE MAGNETOSENSOR BULLA IS THE MUSCLE GENERATOR OF THE MAGNETOSENSOR ORGANS. THE WALLS OF THE MAGNETOSENSOR BULLA ARE FORMED BY FIBROUS NEUROGLIA, AMONG WHICH ARE PASSED INDIVIDUAL NERVE FIBERS ISSUING FROM THE CENTRAL NERVE BUNCH AND INNERVATING THE MUSCLE GENERATOR.

5. THE TERMINAL BRAIN IS LOCATED IN THE ANTERIOR PART OF THE BRAIN CAPSULE AND REPRESENTS LIKE A GROWTH 25 – 30 BY 15 – 20 CM IN SIZE (IN DIFFERENT FORMS OF ALIEN, ITS SIZES VARY) AT THE END OF THE CENTRAL FIBER BUNCH OF THE OLIENCEPHALON BRAIN. OUTSIDE, THE FINAL BRAIN MORE LIKE A WALNUT. ON THE CUT, IT CONSISTS OF PLATES OF GRAY SUBSTANCE SEPARATED BY THIN LAYERS OF CONDUCTIVE FIBERS. THE TERMINAL BRAIN IS RESPONSIBLE FOR ORGANIZING THE MOST COMPLEX FORMS OF ADAPTIVE BEHAVIOR, INCLUDING SPEECH AND INTELLECTUAL ACTIVITY.

SENSORS:

ALIENS HAVE THE FOLLOWING TYPES OF RECEPTORS:

1) DISTANT EXTERORCEPTORS GROUPED INTO SPECIAL SENSE ORGANS – AN ORGAN OF VISION (SIMPLE EYES IN MOST FORMS OF WORKERS AND CERTAIN TYPES OF SOLDIERS AND MORE OR LESS DEVELOPED FACETIES IN OTHERS CASTES), ORGANS OF TOUCH AND AIR PRESSURE (ANTENNAS), CEREBRAL MAGNETOSENSOR ORGANS – A PARTICULAR SUBSTITUTE OF VIEW, USED ALSO FOR RECORDING, EMITION AND EXCHANGE OF INFORMATION IN THREE-DIMENSIONAL DYNAMIC IMAGES (SEE ALIEN LANGUAGE);

2) DISTANT EXTERORECEPTORS, NOT GROUPED INTO SENSORS SCATTERED OVER THE BODY BUBBLE SENSILLA RESPONSIBLE FOR AUDIO AND VIBRATION SENSITIVITY;

3) CONTACT EXTERORECEPTORS – HAIR SENSILLA AND FREE NERVE ENDINGS LOCALIZED ON THE SURFACE AND IN THE THICKNESS OF EXOSKELETAL STRUCTURES, RESPONSIBLE FOR THE PERCEPTION OF PRESSURE, STRETCH AND TEMPERATURE OF THE EXTERNAL ENVIRONMENT;

4) SOMATIC INTERORECEPTERS – FREE AND ENCAPSULATED NERVE TERMINATIONS IN THE THICKNESS OF THE MUSCLE COMPONENT OF THE MYOEPITHELIUM, FORMING THE BODY COVERING, LUNGS, PSEUDOSYRINX AND LIMB, RESPONSIBLE FOR THE PERCEPTION OF MUSCLE CONTRACTIONS AND SENSITIVITY;

5) VISCERAL INTERIORS – FREE AND ENCAPSULATED NERVE ENDINGS, LOCALIZED IN THE WALLS OF VESSELS AND HOLLOW ORGANS, AS WELL AS THE PROCESS OF CELLS OF THE CENTRAL GRAY KERNEL OF VISCERAL BRAIN, RESPONSIBLE FOR THE PERCEPTION OF THE CONCENTRATION OF BASIC METABOLITES AND BLOOD GASES, AS WELL AS THE TONE OF THE WALLS AND MOTOR VESSELS AND HOLLOW ORGANS.

IN BASE OF VISION OF A SMALL WORKER IS A SIMPLE EYE CONSISTING OF:

1) COVER LENS – EXOSKELETAL LENTIFYING FORMATION WITH HIGH TRANSPARENCY;

2) VITRIC MASS – A DENSE VISCOUS LIQUID, ON THE SURFACE OF WHICH THE COVER LENS IS;

3) A SENSOR PLATE CONSISTING OF 30-40 LIGHT-SENSITIVE NEUROEPITHELIOCYTES AND 50-60 SECRETORY CELLS RESPONSIBLE FOR THE PRODUCTION AND RESORPTION OF THE VITRIC MASS, LOCATED BETWEEN THEM. THE EYE IS INCLUDED IN A CONNECTIVE TISSUE CAPSULE, THE BASIC SUBSTANCE OF WHICH IS PAINTED IN A DARK COLOR. PROCESSES OF NEUROEPITHELIOCYTES PERFORM THE CAPSULE UNDERING THE EYE THE MUSCULAR COMPONENT OF THE MYOEPITHELIA AND GO TO THE SPELLITED PROCESS, ASSEMBLY INTO THE OCPITAL NERVE.

<u>THE CHEMORECEPTORY ORGANS ARE ANTENNULES</u>. IT IS AN PROCESS OF LOPOPHORUS OSKELENET. THE BASIS OF THE ANTENNULE IS MADE UP BY A HOLLOW ROD, FROM WHICH THE BRANCHES OF THE FIRST, SECOND AND THIRD ORDER DEPART. ON THE SURFACE OF THE BRANCHES OF THE THIRD ORDER (NON-EXORPETE BRANCHES, BUT ARE MIOEPITHELIAL GROWTHS THAT ARE ABLE TO HIDE IN THE SECOND-ORDER BRANCHES) ARE BEADED CUPULS. THEY ARE

ACCUMULATIONS OF NEUROEPITHELIAL CELLS COVERED WITH «CAPS» FROM VISCOUS MUCUS.

PROCESSES OF NEUROEPITHELIAL CELLS PASS IN THE CAVITIES OF THE BRANCHES AND THE ROD, MERGING AT ITS BASE INTO THE ANTENNULAR NERVE. THE BASE OF THE ROD PUSHES DEEPLY INTO THE MUSCLE COMPONENT OF THE LOPOPHORUS MYOEPITHELIUM, SEVERAL MUSCLE BUNKS OF THE RADIAL DIRECTION ARE ATTACHED TO IT, IMPLEMENTING THE MOVEMENTS OF THE ANTENNULE.

THE CEREBRAL MAGNETOSENSOR APPARATUS IS THE MAIN ORGAN OF THE DISTANT RECEPTION OF A SMALL WORKER. THE PRINCIPLE OF WORK OF THIS SENSOR ORGANS IS REGISTRATION OF A VARIABLE MAGNETIC FIELD OF A COMPLEX CONFIGURATION GENERATED BY THE DEVICE ITSELF, UNDER THE INFLUENCE OF SURROUNDING BODIES. THE MAIN DEPARTMENTS OF THE CEREBRAL MAGNETOSENSOR APPARATUS ARE THE MUSCLE GENERATOR, CREATING AN ELECTROMAGNETIC PULSE OF A COMPLEX CONFIGURATION, AND THE OLIENCTPHALON SENSOR NUCLEI, RECEIVING THE PERTURBATION OF THE VARIANT ELECTROMAGNETIC FIELD UNDER THE ACTION. RECONSTRUCTION OF A THREE-DIMENSIONAL PICTURE OF THE ENVIRONMENT IS MOSTLY IN THE FINAL BRAIN.

THE MUSCLE GENERATOR IS A COLLECTION OF 12-15 LOOSE PROCESSED GLOBULAR FORMATIONS SUSPENDED IN THE MAGNETOLYMPH OF THE MAGNETOSENSOR BULLA. EACH OF THESE FORMATIONS – GENERATING MODULES – CONSISTS OF A DENSE GLOBULAR BODY, SURROUNDED BY A CLOUD OF COBELL-LIKE PROCESSES CONNECTING THE BODY WITH OTHERS. IN THE CENTER OF THE BODY IS A SMALL LONG CRYSTAL OF IRON CARBONATE. HISTOLOGICALLY GENERATING MODULES ARE CONSTRUCTED FROM A SPECIAL EXCITABLE TISSUE SIMULTANEOUSLY SIMILAR TO NERVOUS TISSUE

AUDIO AND VIBRATION RECEPTOR BUBBLE SENSILLA – IS A HYALINE-SHAPED CAVITY IN THE EXOSKELETON FILLED WITH AIR. THE OUTER WALL OF THE CAVITY IS THIN AND UNDER THE INFLUENCE OF VIBRATIONS OF THE RESONANT FREQUENCY COMES INTO MOTION. THE INTERNAL WALL OF THE CAVITY CONSISTS OF MORE SOFT CHITIN PERCEPTED WITH NERVE ENDINGS. RESONANT FREQUENCIES ARE DIFFERENT FOR SENSILLA LOCATED IN DIFFERENT PARTS OF THE BODY, WHICH PROVIDES SOUND AND VIBRATION PERCEPTION IN A WIDE FREQUENCY RANGE. THE TOTAL NUMBER OF BUBBLE SENSILLA ON THE BODY OF A SMALL WORKER IS MORE THAN 27,000.

THE ENDOCRINE GLANDS OF ALIENS ARE REPRESENTED BY: 1) NEUROSECRETORY ELEMENTS OF THE CENTRAL GRAY NUCLEUS OF THE

VISCERAL BRAIN; 2) SUBMACULAR ORGAN; 3) PARAENTERAL BODIES; 4) INTRAMESENTERIAL BODIES; 5) PARAPHARYNGEAL BODIES; 6) ENDOCRINE ELEMENTS OF THE GENERAL GLANDS.

THE SUBMACULAR ORGAN IS LOCATED ON THE ANTERIOR SURFACE OF THE SPEAR-SHAPED PROCESS IN THE UPPER PART OF THE LOPHOPHORE. IT IS A SHAPED FORMATION OF A PALE BLUE COLOR OF SOFT CONSISTENCY, WEIGHT ABOUT 5 G, INCLUDED IN A COMMON CONNECTIVE TISSUE SHELL WITH ONE OF THE OCULAR NERVES. PART OF THE FIBERS OF THE EYE NERVE PENETRATION INSIDE THE SUBMACULAR ORGAN. THE SUBMACULAR ORGA SECRETS THREE POLYISOPRENE COMPOUNDS (KATARGIN, ANABOLIN AND PHOTOTROPIN) RESPONSIBLE FOR THE INDUCTION OF MOLTING IN LARVIES AND REGULATION OF THE DAILY RHYTHM OF ANABOLIC PROCESSES IN ADULTS.

PARAENTERAL BODIES ARE SMALL GLANROUS FORMATIONS IN SIZE FROM A PEA TO A MILLET GRAIN, LOCATED INSIDE THE CAPSULE OF THE GLANDS OF THE SMALL INTESTINE. PARAENTERAL BODIES SECRETE MORE THAN 30 BIOLOGICALLY ACTIVE PEPTIDES, SUPPORTING THE CONCENTRATION OF MAIN METABOLITES IN THE BLOOD AND ALSO AFFECTING INTESTINAL MOTORITY AND SECRETION.

INTRAMESENTERIAL BODIES ARE SPONGEOUS FORMATIONS OF BLUE-GREEN WEIGHT 5-20 G, LOCATED BETWEEN THE LEAVES OF THE COMMON MESENTERI FROM THE STOMACH TO THE COLON, AND ALSO IN THE CAPSULE OF THE GREEN GLAND. THE NUMBER OF INTRAMESENTIAL BODIES IN AN ADULT SMALL WORKER IS FROM 4 TO 7 PCS. INTRAMESENTERIAL BODIES ARE FORMED BY VESSEL PLEXES. THESE VESSELES DEPART FROM THE VISCERAL BRANCHES OF THE ARTERIAL SEGMENT OF THE PARACHORDAL RING. THEY FLOW INTO THE VENOUS SEGMENT OF THE PARACHORDAL RING NEAR THE EXIT OF THE VENOUS DUCT. VESSELS OF THE INTRAMESENTERIAL BODIES ARE CONNECTED TO THE SYSTEMIC BLOOD CIRCULATION, BUT CAN BE OVERLAPPED UNDER THE INFLUENCE OF SPECIFIC ACTIVE PEPTIDES, ACTING SMOOTH MUSCLE SPHINCTERS. THESE PHYSIOLOGICALLY ACTIVE PEPTIDES ARE PRODUCT OF ALERTASE CLEAVING OF SOME BLOOD PROTEINS. ALERTASE IS SECRETED BY THE GLANDS LOCATED IN ADVENTION OF THE VESSELS. UNDER THE IMPACT OF DAMAGED FACTORS ON THE ORGANISM OF THE SMALL WORKING SPINCTERS OF THE VESSELS OF THESE GLANDS RELAX, AND THE ACTIVE PEPTIDES LEFT INTO THE BLOOD. THEY CAUSE SPECIFIC REACTIONS TO DAMAGES: CHANGES IN HEART RATE, VASCULAR TONIS, ELECTRICAL BRAIN ACTIVITY, ENERGY PRODUCTION LEVEL, ETC.

PARAPHARYNGEAL BODIES – PAIR SHAPED GRAY-YELLOW FORMATIONS WEIGHTING ABOUT 7 G, LOCALIZED ON THE OUTER SURFACE OF THE

PHARYNGEATER AROUND LUNG DRAINAGE DUCTS. PARAPHARYNGEAL BODIES SECRET ACTIVE THERMOFORM AMINE, STIMULATING THE HEAT-PRODUCT OF THE FAT CUSHION CELLS WHILE REDUCING THE AMBIENT TEMPERATURE, AND A PEPTIDE, CHANGING THE VOLUME AND CONSISTENCY OF THE GREEN GLAND SECRETION. DEPENDING OF TEMPERATURE AND HUMIDITY THIS PEPTID CHANGE ELSE PULMONARY MUCUS.

EXOCRINE GLANDS OF ALIEN ARE EXTREMELY DIFFERENT IN DIFFERENT FORMS. SMALL WORKER HAS:

1) INTRASYRINGEAL GLANDS – ABOUT 1500 MICROSCOPIC GLANDS LOCATED ON THE SURFACE OF PSEUDOSYRINX TRABECULS. INTRASYRINGEAL GLANDS ARE RESPONSIBLE FOR THE PHEROMONE COMPONENT OF ALIEN'S SPEECH.

2) SUBPHARINGEAL GLAND – A LARGE LOBE FORMATION LOCATED UNDER THE LOWER WALL OF THE PHARYNGEAL. THE EXECUTIVE DUCT OF THE SUBPHARYNGEAL GLAND OPEN AT THE BASIS OF THE GRABING ORGANS. THE SUBPHARYNGEAL GLAND SECRETES THE ONLY POLYISOPRENE NATURE PHEROMONE, CAUSING THE STEREOTYPICAL RESPONSE OF FEEDING IN OTHER LUNG WORKERS. SUB-PHARYNGEAL GLAND SECRETION IS STIMULATED WHEN THE BLOOD GLUCOSE CONTENT FALLS BELOW A CERTAIN LEVEL.

3) MICROSCOPIC COXAL AND TENTAL GLANDS LOCALIZED ON THE SURFACE OF THE THIGH AND TENTACLES. THEYR SECRET IS A PARTICULAR MARK OF SMALL WORKERS, WHICH MAKES EASIER TO CARE FOR SYMBIONTS. IN ADDITION, THE CHEMICAL COMPOSITION OF THESE SECRETS IS DIFFERENT IN ALIENS FROM DIFFERENT NESTS, WHICH FACILITATES MUTUAL RECOGNITION. THE SMELL OF COXAL SECRETS OF RESIDENTS OF OTHER NESTS CAUSES A STUPOR STATE IN THE WORKERS AND A STEREOTYPICAL ATTACK REACTION IN THE SOLDIERS.

4) STATUS GLANDS – A SHAPED WHITE FORMATION WEIGHTING ABOUT 50 G, LOCATED SUBMACULARLY ON THE FRONT SURFACE OF VISCEROTHECA. THEYR DUCT ARE OPENED INTO THE GAP BETWEEN THE BRANCHIOPHORE AND THE VISCEROTHECA. THE SMELL OF THEYR SECRET DEFINES THE CAST TYPE OF PERSON

SOME CYTOLOGICAL AND HISTOLOGICAL DATA.

THE STRUCTURE OF ALIEN CELLS IS SIMILAR TO THE STRUCTURE OF TERRESTRIAL EUKARYOTES CELLS, WHICH, APPARENTLY, IS EXPLAINED BY THE PROXIMITY OF METABOLIC PROCESSES. THE BASIC STRUCTURES OF ALIEN CELLS ARE A PLASMATIC MEMBRANE WITH GLYCOCALYX, PROTOPLASMA AND NUCLEUS. MEMBRANE SYSTEMS ASSOCIATED WITH SECRETION PROCESSES, SEVERAL TYPES OF MEMBRANE VESICLES, CYTOSKELETAL STRUCTURES, AND

BELOOXYNTHESIZING NON-MEMBRANE PARTICLES, FUNCTIONALLY SIMILAR TO RIBOSOMES, ARE LOCALIZED IN PROTOPLASMA.

THE MAIN FEATURES OF THE STRUCTURE OF ALIEN CELLS ARE THESE:

1) EXTREMELY LARGE SIZES. MAJORITY OF ALIEN TISSUES (MYOEPITHELIUM OF BODY COVERINGS, LIMB, DIGESTIVE TRACT, NEUROGLIA, TISSUE OF GREEN GLAND, LUNGS AND PSEUDOSYRINX) CONSISTS OF SIMPLASTS – GIANT MULTINUCLEOR FORMATIONS 1 – 5 MM IN SIZE. CAPILLARIES FREQUENTLY PASS INSIDE THE SYMPLAST CYTOPLASMA IN MEMBRANE CHANNELS CONNECTED WITH THE CAPILLARY PLASMA MEMBRANE (MESOCAPILLARIUM). SINGLE-NUCLEAR ALIEN CELLS ARE ON AVERAGE ALSO BIGGER THAN CELLS OF TERRESTRIAL ANIMALS (MEDIUM SIZES – 50 – 100 μM).

2) STRUCTURE OF THE PLASMATIC MEMBRANE: ALL ALIEN CELLS HAVE AN ADDITIONAL LAYER OF SULFOLIPID CAPSULES, SCREENING THE GLYCOCALYX, OVER THE LIPOPROTEID BILAYER AND GLYCOCALYX. THIS LAYER IS INTERRUPTED ONLY AT THE LOCATION OF INTERCELLULAR CONTACTS.

3) THE PRESENCE IN THE SUBMEMBRANE LAYER OF THE CYTOPLASMA OF MYELIN-LIKE BODIES (SUBMEMBRANE GRANULES) THAT ARE A RESERVE OF THE MEMBRANE MATERIAL IN THE EVENT OF DAMAGE TO THE PLASMALEMMAL.

4) LOCALIZATION OF OXIDATIVE PHOSPHORYLATION PROCESSES. ALIENS DO NOT HAVE MITOCHONDRIA, AND THE FINAL STAGES OF OXIDATIVE METABOLISM OF ENERGY SUBSTRATES OCCUR IN SPECIAL PLASMALEMMAL INTERDIGITATIONS – THE SO-CALLED NET BODIES. THE NAME OF THE LATTER IS CONNECTED WITH THE FACT THAT LONG FINGER-LIKE INVAGINATIONS OF THE PLASMALEMMA CAN STRETCH PRACTICALLY THROUGH THE ENTIRE CELL BODY, TO THE ORGANELLES THAT ACTIVELY EXPEND ENERGY, FOR EXAMPLE, TO THE CONTRACTILE FIBRILS. WHEN THE HISTOLOGICAL REACTION TO OXIDATIVE PHOSPHORYLATION ENZYMES IS CARRIED OUT, THE CYTOPLASMA APPEARS TO BE A FILLED NETWORK FROM DARK BLUE THREADS, IN SOME PLACES ATTACHED TO THE PLASMATIC MEMBRANE.

5) FEATURES OF THE ORGANIZATION OF CHROMOSOMES. THE CHEMICAL COMPOSITION OF ALIEN CHROMOSOMES AND THE WAY OF CHROMATIN FOLDING ARE SIMILAR TO THE TERRESTRIAL EUKARYOTES. HOWEVER, IN CONNECTION WITH THE DEVELOPMENT OF ALLOCIDE-DEPENDENT TRANSDUTION, THE KARYOTYPE IS EXTREMELY VARIABLE: THE CHROMOSOMAL SET CONSTANTLY PRESENT SMALL CHROMOSOMAL FRAGMENTS, RING-SHAPED CHROMOSOMES; THE NUMBER, SIZE AND CHARACTER OF DIFFERENTIAL STAINING OF LARGE CHROMOSOMES IS EXTREMELY VARIABLE. IN ACTUALLY, THE KARYOTYPES ARE IDENTICAL IN THE DESCENDANTS OF ONLY ONE PAIR OF UPPER CAST. LOSS OF GENETIC MATERIAL IS PREVENTED BEACOSE SPINDLE FIBERS CAN BE HAVING

MANY ATTACHEMENT LOCUSES ON CHROMATIN. IN ADDITION, ONLY UPPER CASTS ARE DIPLOID, THE OTHER ARE HAPLOID.

ALL ALIEN ORGANS ARE CONSTRUCTED FROM THREE TYPES OF TISSUE: MYOEPITHELIAL, CONNECTIVE AND NERVOUS.

THE MYOEPITHELIAL TISSUE FORMS BODY COVERS, LIMB, LUNGS, PSEUDOSYRINX, DIGESTIVE TRACT, GREEN GLAND, PARAENTERAL BODIES, PARAPHARYNGEAL BODIES, AND EXOCRINE GLANDS. TYPYCAL SINGLE-CORE CELL OF MYOEPITHELIAL TISSUE (LUNGS ARE BUILT FROM THESE ELEMENRS) IS REPRESENTED A CUBIC BODY WITH LIGHT CYTOPLASM, INCLUDED AN OVAL KERNEL (EPITHELIAL COMPONENT) AND THE BASAL AREA OF THE DARK-COLORED CYTOPLASM, FILLED WITH LONGITUDINALLY ORIENTED DENCE PROTOFIBRILS (COMPONENT). THE CAPILLARIES PASS INTRACELLULARLY WITHIN THE BASAL ARIA. THE EPITHELIAL COMPONENT IN THE LUNGS HAS CILIAS. HOWEVER, THEY MAY BE ABSENT. BOTH THE EPITHELIAL AND MUSCLE COMPONENTS OF PARTICULAR MYOEPITHELIOCYTES CAN FUSE TO FORM SYMPLASTS. HOWEVER, USUALLY THIS HAPPENS WITH THE MUSCLE COMPONENT. DEPENDING ON THE RATIO OF MUSCLE AND EPITHELIAL COMPONENTS.

THERE ARE 5 GENERAS OF MYOEPITHELY:

THE TYPE I MYOEPITHELIUM FORMES THE LIMBS, BODY COVERING AND IS CHARACTERIZED: 1) A FLAT CHITIN-SECRETING EPITHELIAL COMPONENT; 2) A WELL DEVELOPED MUSCLE COMPONENT FORMING A SIMPLAST.

THE TYPE II MYOEPITHELIUM IS CHARACTERIZED BY A FLAT MUSCUS-PROCESSING EPITHELIAL COMPONENT. THE MUSCLE COMPONENT IS SIMILAR TO THAT OF THE TYPE I MYOEPITHELIA. THE MYOEPITHELIUM OF THE II TYPE FORMS THE THROAT, STOMACH AND LARGE INTESTINE.

THE TYPE III MYOEPITHELIUM IS CHARACTERIZED BY A CYLINDRICAL PHAGOCYTED HYDROLASE-SECRETING EPITHELIAL COMPONENT AND FORMES THE SMALL INTESTINE.

THE TYPE IV MYOEPITHELIUM IS CHARACTERIZED BY CUBE CILIATED MUCUS-SECRETING EPITHELIAL COMPONENT AND FORMES LUNGS, PSEUDOSYRINX, GREEN GLAND AND EXOCRINE GLANDS (INTRASYRINGEAL).

THE TYPE V MYOEPITHELIUM IS CHARACTERIZED BY A MUSCLE COMPONENT REDUCED TO SEVERAL MYOPHIBRILES WITH HIGH SECRETORY AND METABOLIC ACTIVITY. CAPILLARIES PASS IN THE CYTOPLASMA OF THE EPITHELIAL COMPONENT. MYOEPITHELIUM OF THE V GENUS FORMS GREEN GLANDS AND EXOCRINE GLANDS.

THE CONNECTIVE TISSUE OF ALIENS IS SIMILAR TO THAT OF TERRESTRIAL VERTEBRATES AND IS FORMED BY SEVERAL TYPES OF SINGLE-NUCLEAR CELLS

AND A MORE OR LESS DEVELOPED INTERCELLULAR SUBSTANCE. CONNECTUAL STRUCTURES INCLUDE A TOTAL MEMBRANE, HEART, VESSELS, A FAT BODY, THE MESOTHELIAL LINER OF BRANCHIOTHECE, SIRINGOTHECE AND VISCEROTHECA, THE SHELL OF PERIPHERAL NERVES, A MEMBRANE, WHICH CLOSES THE TRIANGULAR SLOT OF BRAIN CAPSULES, AND CAPSULES OF SIMPLE EYES. SPECIAL TYPES OF CONNECTIVE TISSUE ARE BLOOD AND ENDOSKELETON OSSEOID TISSUE.

THE CELLULAR ELEMENTS OF THE CONNECTIVE TISSUE OF ALIENS BELONG TO TWO MAIN TYPES – PROTECTIVE AND EPITHELIOID.

PROCESSING CELLS ARE CHARACTERIZED BY THE PRESENCE OF NUMEROUS FILLED CYTOPLASMATIC PROCESSES WHICH PENETRATE DEEPLY INTO THE INTERCELLULAR SUBSTANCE, PERFORMING THE FUNCTION OF ITS SECRETION AND RESORPTION, AND CONTACT WITH SIMILAR PROCESSES. THESE INCLUDE:

1. FIBROCYTE – A CELL OF MEDIUM SIZE WITH LIGHT CYTOPLASMA, A LARGE ROUND NUCLEUS AND NUMEROUS VESICLES CONTAINING HYDROLYTIC ENZYMES. FIBROCYTES ARE RESPONSIBLE FOR BOTH SECRETION AND RESORPTION OF THE GROUND SUBSTANCE OF CONNECTIVE TISSUE FORMATIONS (ALIENS DO NOT HAVE SPECIALIZED MACROPHAGES).

2. SMOOTH MUSCLE CELL IS MORPHOLOGICALLY SIMILAR TO FIBROCYTE BUT HAS A DARK CYTOPLASMA DUE TO A LARGE NUMBER OF RANDOMLY INTERNATED MYOPIBRILS. PROCESSES OF NEIGHBOR SMOOTH MUSCLE CELLS CONTACT BETWEEN THEM, FORMING ELECTRIC SYNAPSE. SMOOTH MUSCLE CELLS FORM THE «HEART» OF FOREIGN AND THE MIDDLE LAYER OF THEIR VESSEL WALLS (OUTSIDE THE VESSELS ARE COVERED WITH A THIN LAYER OF FIBROCYTES AND INTERCELLULAR SUBSTANCE, AND INSIDE WITH ENDOMESOTHELIA).

3. CHORDOBLAST IS AN ANALOGUE OF FIBROCYTE FOR ENDOSKELETON OSSEOID TISSUE AND DIFFERS FROM THE LAST IN AN ASYMMETRIC LOCATION OF PROCESSES (CELL SURFACE DIRECTIONAL TO THE OUTER EDGE OF ENDOSKELETON STRUCTURE DOES NOT HAVE PROCESSES).

EPITHELIOID CELLS ARE CHARACTERIZED WITH A FLAT SHAPE AND A SMOOTH SURFACE. THESE INCLUDE:

1. ENDOMESOTHELIOCYTE – A FLAT CELL WITH A LIGHT CYTOPLASMA AND A ROB-SHAPED NUCLEUS. ENDOMESOTHELIOCYTES FORM THE OUTER LAYER OF THE GENERAL MESENTERY, THE INTERNAL LAYER OF THE WALLS OF THE SYRINGOTEKA, BRANCHIOTHECA AND VISCEROTHECA, THE INTERNAL LAYER OF THE VESSEL WALLS AND THE SINGLE LAYER OF THE CAPILLARY WALLS.

2. FAT BODY CELL – A FLAT CELL WITH AN ECCENTRICLY POSITIONED BEAN-SHAPED NUCLEUS AND PLENTY OF CYTOPLASMATIC VACUOLES, CAPABLE OF

AMEBOID MOVEMENT. CYTOPLASMATIC VACUOLES CONTAIN LIPID DROPS, AMINOPECTIN GRANULES (RESERVE POLYSACCHARIDE, SIMILAR TO GLYCOGEN) AND A CONCENTRATED SOLUTION OF AMINO ACIDS. FAT PAD CELLS ARE ABLE TO PENETRATE BOTH THROUGH THE CAPILLARY WALL INTO THE BLOOD COURSE AND FROM THE BLOOD COURSE INTO THE TISSUE. NORMALLY, 90% OF THESE CELLULAR ELEMENTS ARE CONCENTRATED IN THE FAT BODY, AND 10% ARE IN THE BLOOD FLOW. THE FUNCTION OF THE CELLS OF THE FAT DODY TRANSPORT LIPIDS FROM THE DIGESTIVE TRACT TO THE EFFECTIVE ORGANS, DEPOSIT THE BASIC NUTRIENTS AND MAINTAIN CONSTANTITY FROM THE CONCENTRATION IN THE BLOOD. UNDER CONDITIONS OF OVERFEEDING OR DISFUNCTION OF PARAENTERAL BODIES, FAT BODY CELLS CAN ACCUMULATE IN VESSEL WALLS AND INTERNAL ORGANS.

INTERCELLULAR SUBSTANCE CONNECTING TISSUE CONSISTS OF SEVERAL TYPES OF STRUCTURAL POLYSACCHARIDES AND FIBRILLARY KERATIN-LIKE PROTEIN. INTERCELLULAR SUBSTANCE OF ENDOSKELETON OSSEOID TISSUE IS DIFFERENT IN THE INCLUSION OF CRYSTALS OF ARAGONITE AND IRON CARBONATE, AS WELL AS A LARGE AMOUNT OF SULFATED POLYSACCHARIDES. INTERCELLULAR SUBSTANCE OF A SIMPLE EYE CAPSULE CONTAINS GRAINS OF DARK PIGMENT OF UNKNOWN CHEMICAL NATURE.

CELLS AND BLOOD CHEMICAL COMPONENTS ARE DESCRIBED IN THE ALIEN METABOLISM AND PHYSIOLOGY SECTION.

NERVOUS TISSUE CONSISTS OF CELLS OF THREE TYPES OF NEURONS, PERISOMATIC GLIA AND GLIA WITH LIPID INCLUSIONS.

ALIEN NEURONS ARE SIMILAR TO EARTH VERTEBRATE NEURONS. DIFFERENCES IN THE STRUCTURE ARE OF ULTRASTRUCTURAL CHARACTER AND REDUCED TO: 1) THE PRESENCE ON THE BODY AND PROCESSES OF THE NERVE PULSE CONDUCTION BANDS AND THE NETWORK BODIES AND SUBMEMBRANE GRANULES DESERVED IN CONNECTION WITH THIS; 2) THE ABSENCE OF SPECIALIZED SYNAPTIC FORMATIONS MOST NEURONS ARE CONNECTED BY ELECTRIC SYNAPSE LOCATED AT THE PLACES OF CONTACT OF THE PROCESSES OF TWO NEURONS OR THE PROCESS OF A NEURON AND PERISOMATIC GLIA.

PERISOMATIC GLIA ENVELOPES THE BODIES OF NEURONS AND THE INITIAL SEGMENTS OF THEIR PROCESSES. IT IS REPRESENTED BY GIANT SHAPE SIMPLASTS, INTO THE CYTOPLASMA OF WHICH FROM 1 TO 5-6 NEURONS AND THE CAPILLARIES FEEDING THEM ARE INSIDE. AT THE LOCATION OF THE CONDUCT BANDS BETWEEN THE PLASMALEMMAS OF THE PERISOMATIC GLIOCYTE AND THE NEURON, THERE ARE GAPS UP TO 10 μM WIDE, FILLED WITH THE FIBRILLAR MATRIX.

GLIA WITH LIPID INCLUSIONS COVERS THE BANDS OF NEURON PROCESSES. IT CONSISTS OF SINGLE-NUCLEAR CELLS OF OBLONG SHAPED ABOUT 100 MKM LONG WITH A ROB-SHAPED ECCENTRICLY POSITIONED NUCLEUS. PART OF THE GLIOCYTE CYTOPLASMA ADJACENT TO THE NEURON PROCESS IS OCCUPIED WITH A GIANT VACUOL FILLED WITH A MIXTURE OF SULFOLIPIDS. INDIVIDUAL GLIOCYTES DURING THE PERFORMANCE STRIP ARE SEPARATED BY SLICES 10 -15 μM WIDTH. THE ROLE OF GLIA WITH LIPID INCLUSIONS – PARTIAL ELECTRICAL ISOLATION OF THE MEMBRANE OF CONDUCT BANDS LEADING TO ACCELERATION OF NERVE IMPULSE CONDUCTION

METABOLISM AND PHYSIOLOGY OF ALIENS

THE BODY OF ALIENS IS CONSTRUCTED PRACTICALLY FROM THE SAME «BUILDING BLOCKS» AS THE BODIES OF ABOVEGROUND ANIMALS. THE DIFFERENCES CONCERN, MAINLY, TO THE STRUCTURE OF MUCOUS AND CONNECTIVE TISSUE POLYSACCHARIDES AND, TO A CERTAIN DEGREE, TO THE STRUCTURE OF NUCLEIC ACIDS. ALIENS ARE AEROBES, HOWEVER, THEIR NEED FOR OXYGEN IS SIGNIFICANTLY LOWER THAN ABOVEGROUND ANIMALS OF THE SAME MASS, WHICH IS ASSOCIATED WITH THE FEATURES OF COMPARTMENTALIZATION OF OXIDATIVE PHOSPHORYLATION. THE OPTIMAL COMPOSITION OF THE GAS MIXTURE FOR ALL FORMS OF ALIENS IS 13% OXYGEN, 2% CARBON DIOXIDE AND 85% INERT GAS AT A PRESSURE OF 1 ATA AND A RELATIVE HUMIDITY OF 90 – 100%, WHICH CORRESPONDS TO THE COMPOSITION OF AIR INSIDE BUILDINGS. WHEN BREATHING ATMOSPHERIC AIR CONTAINING 21% OXYGEN FOR MORE THAN 12 HOURS, THE TOXIC EFFECT OF OXYGEN BEGINS TO BE MANIFESTED, MORE PROGRESS IN UPPER CASTS. A REDUCTION OF THE HUMIDITY OF THE GAS MIXTURE TO 40% AND LESS FOR A TERM OF MORE THAN SEVERAL HOURS LEADS TO SPEECH AND RESPIRATORY DISTURBANCES. CHANGES IN THE CONCENTRATION OF CARBON DIOXIDE WITHIN A WIDE RANGE (0.1 – 7%) DO NOT LEAD TO ANY HARMFUL CONSEQUENCES.

A FEATURE OF THE CARBOHYDRATE AND LIPID METABOLISM OF ALIENS IS THE ABILITY OF THE LATEST TO SYNTHESIZE OXALOCETIC ACID FROM TWO-CARBON FRAGMENTS

THE FINAL PRODUCT OF NITROGEN METABOLISM IN ALIENS IS URIC ACID, SYNTHETIZED BY THE CELLS OF THE GREEN GLAND AND EXHAUSTED WITH THE SECRET OF THE LATEST. ALIENS ARE ALSO ABLE TO RESYNTHESIZE PURINE BASES FROM URIC ACID.

AN ESSENTIAL AIMNO ACID FOR ALIENS IS PHENYLALANINE.

NO KNOWN VITAMINS ARE NECESSARY FOR ALIENS, BECAUSE THE LATEST COFERMENTS HAVE NOTHING TO DO WITH THE COFERMENTS OF ABOVEGROUND ANIMALS. THESE ARE ORGANOMETALLIC COMPOUNDS OF PARPHYRIN OR CORRIN NATURE, MOST OFTENLY STRONGLY BOUND TO THE PROTEIN, AND SEVERAL QUINOID COMPOUNDS WITH LONG SIDE CHAINS. THE ORGANIC PART OF THE COFERMENTS IS SYNTHESIZED IN THE ALIEN'S BODY FROM AMINO ACIDS AND KETONE BODIES, AND THE NECESSARY METALS, MAINLY COPPER, MANGANESE AND ZINC, ARE SUPPLIED BY THE EATER'S MICROFLORA OR ARE OBTAINED BY CANNIBALISM.

THUS, FOR OTHER PEOPLE'S INDISPENSABLE NUTRIENTS, THE SOURCE OF ORGANIC CARBON IS (THEY CAN BE PROTEINS, FATS, MONO- OR POLYSACCHARIDES), SOURCE OF AMINO GROUPS (ANY PROTEIN OF PLANT OR ANIMAL ORIGIN), THE AMINO ACID PHENYLALANINE, COPPER, MANGANESE, ZINC AND IRON IONS. THE END PRODUCTS OF METABOLISM ARE CARBON DIOXIDE, METABOLIC WATER AND URIC ACID.

PHYSIOLOGY OF WATER-ELECTROLYTE METABOLISM AND MEMBRANE PROCESSES

THE MAIN INTRACELLULAR CATION OF THE ALIENS IS POTASSIUM BOUND WITH HIGH-MOLECULAR OSMOTICALLY LOW-ACTIVE POLYANIONS (MAJORLY PROTEINS). CALCIUM AND LOW-MOLECULAR ANIONS – CHLORIDE AND BICARBONATE – ARE IN INTRACELLULAR LIQUID IN TRACE QUANTITIES. THE EXTRACELLULAR FLUID OF THE ALIENS CONTAINS POTASSIUM AND CALCIUM CHLORIDES AND POTASSIUM BICARBONATE, HOWEVER, THE MAIN PART OF THE OSMOLARITY

EXTRACELLULAR SECTOR CREATES NON-ELECTROLYTE – INTERMEDIATE PRODUCT OF NITROGEN METABOLISM DIMETHYLAMINOXIDE. SODIUM IS NOT DETERMINATED IN THE EXTRACELLULAR OR IN THE INTRACELLULAR LIQUID OF ALIENS.

THEREFORE, THE IONIC COMPOSITION OF THE EXTRACELLULAR AND INTRACELLULAR LIQUID OF ALIENS HAS THE FOLLOWING FORM:

THE OSMOLARITY OF THE EXTRACELLULAR AND INTRACELLULAR ALIEN FLUID IS ABOUT 800 MOSM/L.

THEREFORE, THE EXTRACELLULAR AND INTRACELLULAR SECTOR OF ALIENS, UNLIKE THE SITUATION OBSERVED IN TERRESTRIAL ANIMALS, IS ISOOSMORAL. BECAUSE THE ALIEN CELLS ARE COVERED WITH AN INSULATING LAYER OF SULFOLIPID GLOBULES, THEIR CYTOLEMMA DOES NOT HAVE PORE PORES TO PROVIDE ANY SIGNIFICANT NON-SPECIFIC ION LEAKAGE. CONSEQUENTLY, UNLIKE THE CELLS OF TERRESTRIAL ANIMALS, ALIEN CELLS IN

THE STATE OF REST DO NOT NEED ENERGY DEPENDENT ION TRANSPORT SYSTEMS TO MAINTAIN THE IONIC COMPOSITION AND OSMOLARITY OF THE INTRACELLULAR SECTOR. THIS MAKES CELLS AND EVEN WHOLE ALIENS HIGHLY RESISTANT TO FREEZING AND DRYING AND ALSO CAUSES SOME FEATURES OF CELL DEATH PROCESSES.

THE MEMBRANE OF ALIEN CELLS HAS FOUR TYPES OF TRANSPORTATION SYSTEMS:

1) ELECTRONIC TRANSPORT SYSTEM OF ELECTRONS AND PROTONS LOCALIZED IN THE REGION OF NET BODIES. THE DONOR OF ELECTRONS IS A REDUCED FORM OF ONE OF THE QUINOID COFERMENTS, AND THE FINAL ACCEPTOR IS OXYGEN. IN THE TRANSFER OF ONE PAIR OF ELECTRONS FROM THE QUINOID COENZYME TO OXYGEN INTO THE EXTRACELLULAR SPACE, FIVE PAIRS OF PROTONS ARE DISPLAYED. PROTONS ARE RETURNED INSIDE THE CELL THROUGH A SPECIAL ION CHANNEL (PROTON-DEPENDENT GTPSYNTHETASE GENERATING GUANOSINTRIPHOSPHATE – THE MAIN MACROERGIC COMPOUND OF ALIENS), OR WHEN THE SYMPORT AND ANTIPORT SYSTEMS WORK (SEE BELOW). INTRACELLULAR PROTONS COMBINE WITH SUPEROXIDE ANION TO FORM METABOLIC WATER.

2) SYSTEMS OF PROTON-DEPENDENT ANTIPORT – INTEGRATE PROTEINS LOCALIZED IN THE MEMBRANE OF NET BODIES INTRODUCING A PROTON INTO THE CELL IN EXCHANGE FOR THE OUTPUT OF ANY ION. THERE ARE AT LEAST 4 PROTEINS OF THIS TYPE – POTASSIUM-PROTON, CALCIUM-PROTON, BICARBONATE-PROTON AND AMMONIUM-PROTON EXCHANGER.

3) PROTON-DEPENDENT SYMPORT SYSTEMS – INTEGRAL PROTEINS LOCALIZED IN THE MEMBRANE BODIES MEMBRANE, PROVIDING THE JOINT INCOME OF THE PROTON AND ANY METABOLITE INTO THE CELL. THESE SYSTEMS PROVIDE SUGAR, AMINO ACIDS, KETO ACIDS, NITROGENIC BASES AND PHOSPHATE ION INTO THE CELL.

4) IONIC CHANNELS – INTEGRAL PROTEINS WITH A TERTIARY STRUCTURE FORMING A PORE TRANSMITTING ION, READINGS RESPONSIBLE FOR THE ION SELECTIVITY OF THE CHANNEL, AND A GATE MECHANISM, SUPPORTING THE PORE IN THE OPEN OR CLOSED STATE

ALIEN CELLS HAVE 6 TYPES OF MEMBRANE CHANNELS:

1) POTASSIUM REST CHANNEL WITHOUT A GATE MECHANISM;

2) ANION CHANNEL (AVAILABLE IN CELLS OF ALL TYPES);

3) POTENTIAL-DEPENDENT POTASSIUM CHANNEL WITH A GATE MECHANISM LOCALIZED ON THE OUTER SIDE OF THE MEMBRANE (CHARACTERISTIC ONLY

FOR EXCITABLE CELLS – NEURONS, MYOEPITHELIA TYPES I-IV AND SMOOTH MYOCYTES);

4) CALCIUM-DEPENDENT POTASSIUM CHANNEL WITH A GATE MECHANISM LOCALIZED ON THE INNER SIDE OF THE MEMBRANE (CHARACTERISTIC EXCLUSIVELY FOR RECEPTOR STRUCTURES);

5) POTENTIAL-DEPENDENT CALCIUM CHANNEL (CHARACTERISTIC ONLY FOR TYPES I-IV MYOEPITHELIA AND SMOOTH MYOCYTES); 6) CHEMOSENSITIVE CALCIUM CHANNEL (HAVE IN CELLS OF ALL TYPES, OPEN IN RESPONSE TO VARIOUS CHEMICAL STIMULES, MEDIATING, IN PARTICULAR, THE ACTION OF HORMONES, BUT ESPECIALLY CHARACTERISTIC FOR RECEPTOR STRUCTURES).

WITH THE EXCEPTION OF THE ELECTRON TRAFFIC CIRCUIT, ALL SPECIFIED TRANSPORT SYSTEMS HAVE A CLEAR PROPERTY THAT ARISES AT THE MOMENT OF CONTACT WITH THE SUBSTRATE, CONFORMATIONAL CHANGES CAUSE PERTURBATIONS OF THE MAGNETIC FIELD OF THE MEMBRANE LEADING TO THE OCCURRENCE OF PORES IN THE SHIELDING LAYER OF SULPHOLYPID GLOBULE. AFTER THE CATALYTIC ACT IS COMPLETE, THE TIME IS IMMEDIATELY CLOSED. OBVIOUSLY, THIS PREVENTS ACCUMULATION OF TRANSPORTED METABOLITES IN THE SPACE BETWEEN THE MEMBRANE AND THE SHIELDING LAYER.

THE OPERATION OF THE IONTRANSPORT SYSTEMS CREATES A RESTING MEMBRANE POTENTIAL EQUAL TO APPROXIMATELY -90 MV. UNLIKE TO TERRESTRIAL VERTEBRATES, THE MEMBRANE POTENTIAL OF CELLS OF DIFFERENT TYPES IS PRACTICALLY NO DIFFERENT IN ALIENS.

<u>THE ELECTRICAL RESPONSES</u> OF ALIEN CELLS ARE OF TWO TYPES:

1) LOCAL GRADUAL POTENTIALS CAUSED BY THE OPENING OF CHEMOSENSITIVE CALCIUM CHANNELS INTO CELLS OF ALL TYPES AND, IN PARTICULAR, TO RECEPTOR STRUCTURES;

2) ACTION POTENTIALS ASSOCIATED WITH PHASE CHANGES IN THE CONDUCTIVITY OF POTENTIAL-DEPENDENT (IN MYOEPITHELIA TYPES I-IV AND SMOOTH MYOCYTES) AND CALCIUM-DEPENDENT (IN RECEPTOR STRUCTURES) POTASSIUM CHANNELS, INDIVIDUAL ONLY EXCITABLE.

<u>THE MECHANISM OF GENERATION OF LOCAL GRADUAL POTENTIAL</u> IS SUCH: IN RESPONSE TO A CHEMICAL STIMULUS (HORMONE OR SECOND INTERMEDIARY OBTAINED DURING THE PRIMARY RECEPTOR ACT), CHEMOSENSITIVE CALCIUM CHANNELS OPEN. INCREASING THE CALCIUM CONDUCTIVITY OF THE MEMBRANE LEADS TO DEPOLARIZATION AND INCREASED INTRACELLULAR CONTENT OF IONIZED CALCIUM, CAUSING MULTIPLE METABOLIC EFFECTS. IN PARTICULAR, IN RECEPTOR STRUCTURES, IT CAUSES CLOSURE OF CALCIUM-DEPENDENT POTASSIUM CHANNELS AND GENERATION OF ACTION POTENTIAL ACCORDING TO THE GENERAL MECHANISM (SEE BELOW).

THE MECHANISM FOR GENERATING OF THE ACTION POTENSIAL IS AS FOLLOWS: UPON REACHING THE CRITICAL LEVEL OF DEPOLARIZATION CAUSED BY THE ARRIVAL OF THE CURRENT LOOP FROM THE NEIGHBORING EXCITABLE CELL THROUGH THE ELECTRICAL SYNAPS, OR THE CLOSURE OF CALCIUM-DEPENDENT POTASSIUM CHANNELS, THE CLOSURE OF POTENTIAL-DEPENDENT POTENTIAL CHANNELS BEGINS, WHICH CAUSES A DECREASE IN THE POTENTIAL CONDUCTIVITY OF THE MEMBRANE FROM THE INSIDE FROM 1 UP TO APPROXIMATELY 0.1. DEPOLARIZATION TO A SPIKE LEVEL OF APPROXIMATELY -32 MV LEADS TO AN AVALANCHE GROWTH IN THE NUMBER OF CLOSED POTASSIUM CHANNELS. AFTER THE SPIKE LEVEL IS REACHED, AN AVALANCHE-LIKE OPENING OF POTENTIAL-DEPENDENT POTASSIUM CHANNELS STARTS, LEADING TO AN INCREASED POTASSIUM CONDUCTIVITY OF THE MEMBRANE AND A RETURN OF THE MEMBRANE POTENTIAL TO A VALUE CLOSEN TO THE RESTING MEMBRANE POTENTIAL. EXCESS POTASSIUM ENTERING THE CELL DURING THE SPIKE TIME THROUGH THE RESTING POTASSIUM CHANNELS, IS REMOVED BY THE POTASSIUM-PROTON EXCHANGER.

THE MECHANISM OF MATING MUSCLE CONTRACTION WITH THE POTENTIAL OF ACTION IS THAT ON THE DOWNWARD PHASE OF THE POTENTIAL OF ACTION TO ACHIEVE THE MEMBRANE POTENTIAL OF THE LEVEL OF APPROXIMATELY -40 MV, POTENTIALLY SENSITIVE CALCIUM CHANNELS ARE OPENED, WITH THE RESULT THAT THE LEVEL OF IONIZED CALCIUM IN THE CELL INCREASES FROM 10-9 TO ABOUT 10-8 M /L. THIS CAUSES THE RELEASE OF ENDOGENOUS CALCIUM FROM THE MEMBRANE VESICLES, AS A RESULT OF WHICH THE LEVEL OF IONIZED CALCIUM IN THE CYTOPLASMA INCREASES TO 10-6 M/L. CALCIUM ACTIVATES ATPASE SYSTEMS OF MYOPIBRILS AND CAUSES THEIR REDUCTION. AT THE END OF THE CONTRACTIONAL ACTION, EXCESS CALCIUM IS REMOVED BY THE CALCIUM-PROTON EXCHANGER.

WATER-ELECTROLYTE EXCHANGE AT THE ORGANIZED LEVEL

IN ALIENS IS DETERMINED BY THE FOLLOWING FACTORS:

1) THE OSMOLARITY OF THE EXTRACELLULAR FLUID IS DETERMINED MOSTLY BY THE CONTENT OF DIMETHYLAMINE OXIDE, HOWEVER, SURPRISING CHANGES IN THE EXTRACELLULAR POTASSIUM CONCENTRATION LEAD TO IMPAIRED TRANSMISSION OF EXCITATION TO THE CNS, MUSCLE CONTRACTION, AND ACTIVITY OF THE RECEPTORS;

2) DEHYDRATION OF ALIEN'S ORGANISM EVEN IN HIGH DEGREES (20 – 30%) DOES NOT LEAD TO SIGNIFICANT NEGATIVE CONSEQUENCES;

3) ESSENTIAL IONS FOR ALIENS ARE CALCIUM, IRON, ZINC, COPPER, MANGANESE AND PHOSPHATE ION.

THE MAIN REGULATORS OF WATER-SALT METABOLISM IN ALIENS ARE THE GREEN GLAND, THE INTESTINE, THE CENTRAL GRAY SUBSTANCE OF THE VISCERAL BRAIN AND THE PARAPHARYNGEAL BODIES. GREEN IRON EXCRETES EXCESS CATIONS IN THE COMPOSITION OF MUCOPOLYSACCHARIDE SECRET AND ALSO SYNTHESIZES DIMETHYLAMINE OXIDE FROM UREA. THROUGH THE INTESTINE (MAJORLY THE COLON), WATER AND ESSENTIAL IONS ARE ABSORBED, AS WELL AS EXECRETION OF PHOSPHATE ION. PARAPHARYNGEAL BODIES RELEASE BIOLOGICALLY ACTIVE PEPTIDES PROVIDING THE FOLLOWING EFFECTS:

1) INCREASED EXECRETION OF POTASSIUM AND ESSENTIAL IONS BY THE GREEN GLAND WITH A DECREASE IN HYDRATION OF THE PULMONARY AND PSEUDOSYRINGEAL MUSCUS (THE LUNGS AND PSEUDOSYRINX ARE THE MAIN SOURCE OF WATER LOSS IN ALIENS);

2) INCREASING THE SYNTHESIS OF DIMETHYLAMINE OXIDE UNDER THE SAME CONDITIONS;

3) INCREASING THE ABSORPTION OF ESSENTIAL IONS (EXCEPT FOR PHOSPHATE) IN THE INTESTINE WITH A DECREASE IN THEIR CONTENT IN THE BODY.

CELLS OF THE CENTRAL GRAY SUBSTANCE OF THE VISCERAL BRAIN HAVE THE FOLLOWING FUNCTIONS:

1) BRAKING THE SYNTHESIS OF DIMETHYLAMINE OXIDE BY THE GREEN IRON WHEN THE OSMOLARITY OF THE EXTRACELLULAR FLUID EXCEEDS OSMOLARITY OF THE INTRACELLULAR FLUID;

2) BRAKES THE ABSORPTION OF ESSENTIAL CATIONS IN THE INTESTINE WHEN INCREASING THEIR CONCENTRATION IN THE BODY; 3) REGULATE THE BALANCE OF PHOSPHATE ION IN THE BODY.

PHYSIOLOGY OF EXTERNAL RESPIRATION.

UNLIKE TERRESTRIAL VERTEBRATES, EXHAUST IS ACTIVE IN ALIENS. THE RESPIRATORY CYCLE IS ADDED FROM THE FOLLOWING COMPONENTS:

1) SYSTOLE OF THE MYOEPITHELIA – SIMULTANEOUS REDUCTION OF THE PULMONARY TRABECULS, CAUSING THE EXPULSION OF OXYGENATED BLOOD FROM THE LUNGS INTO THE PULMONARY VEINS AND AIR FROM THE INTERTRABECULAR SPACES THROUGH THE RESPIRATORY HOLE ON THE TOP OF THE BRANCHIOPHER OUT;

2) MYOEPITHELIAL DIASTOL – RELAXATION OF THE TRABECULA, LEADING TO THE FILLING OF THE INTERTRABECULAR SPACES WITH FRESH AIR, AND THE PULMONARY CAPILLARIES WITH DEOXYGENATED BLOOD;

3) DIFFUSION PAUSE – THE LONGEST PART OF THE RESPIRATORY CYCLE, STARTING AT THE MOMENT OF COMPLETE RELAXATION OF THE TRABECULA AND ENDING WITH THE BEGINNING OF SYSTOLE.

THE DURATION OF THE RESPIRATORY CYCLE VARIES IN WIDE LIMITS AND DEPENDS ON THE OXYGEN CONCENTRATION IN THE INHALED AIR. DURING DIASTOLA AND DIFFUSION PAUSE, THE CILIA OF THE EPITHELIAL COMPONENT ACTIVELY PROPEL THE MUCUS TOWARDS THE DRAINAGE DUCT. THE FREQUENCY OF MYOEPITHELIAL CONTRACTION DUE TO THE GENERAL INNERVATION OF THE LUNGS, LIMB AND INTERMEDIATE MUSCLES IS REGULATED ARBITRARY.

IT SHOULD BE NOTED THAT DUE TO THE LACK OF UPPER AIRWAYS, ALIEN LUNGS ARE VERY SENSITIVE TO THERMAL STRESS AND AIR POLLUTION WITH AEROSOLS.

ALIEN BLOOD – TRANPARANTE OPALESCIENT LIQUID, IT IS DARK BLUE OCSIGENATED AND BLUE-GREEN IN DEOXYGENATED STATE. THERE ARE PLASMA AND BLOOD FORMER ELEMENTS. THE LAST MAKE ABOUT 15% OF THE BLOOD VOLUME. BLOOD PLASMA OF ALIENS IN THE CONTENT OF LOW-MOLECULAR SUBSTANCES IS SIMILAR TO EXTRACELLULAR FLUID. PLASMA FUNCTIONS: 1) GAS TRANSPORT; 2) METABOLITE TRANSPORT; 3) BUFFER; 4) COAGULATION. MOST OF THESE FUNCTIONS ARE MEDIATED BY PLASMA PROTEINS.

GAS TRANSPORT FUNCTION IS MEDIATED BY THE ALIENOCYANINE. IT IS THE CUPROTEIDE CONTAINING FOUR COPPER ATOMS PER ONE MOLECULE, IT CONDITIONS THE BLOOD COLOR OF ALIENS. THE ALIENOCYANIN PROMOTES THE BGAS TRANSFER FUNCTION. IT IS SORBED ON THE SURFACE OF BLOOD FORMED ELEMENTS AND FORMS SUPERAMOLECULAR COMPLEXES WITH PLASMA LIPIDS.

METABOLITE TRANSPORT FUNCTION IS MEDIATED BY 10-20 HIGH MOLECULAR PROTEINS DISSOLVED IN IT, HAVING SPECIFIC BINDING SITES FOR ESSENTIAL IONS, PHOSPHOLIPIDS AND HORMONES OF NON-PEPTIDE NATURE.

BUFFER FUNCTION IS MEDIATED BY BICARBONATIONS AS WELL AS FREE AMINE GROUPS AND CARBOXY GROUPS OF THE PLASMA PROTEIN.

THE COAGULATION FUNCTION OF PLASMA CONSISTS IN THE FORMATION OF A DENSE CLOT IN THE PLACE OF DAMAGE TO THE VASCULAR WALL. IT SHOULD BE NOTED THAT DUE TO THE DENSE STRUCTURE OF THE ALIEN'S BODIES AND TO CHARACTERISTICS OF FOOD PHYSIOLOGY (SEE BELOW), THE VASCULAR WALL MAY BE DAMAGED ONLY WITH SIMULTANEOUS MASSIVE DAMAGE TO THE TISSUES. THIS WAY THE HIGH-SPEED HEMOSTASIS MECHANISMS AT THE MICROCIRCULATORY LEVEL IS NOT NEEDED. SUBMEMBRANE BODIES OF ALL ALIEN CELLS CONTAIN A THERMOSTABLE COAGULATION ACTIVATOR RELEASED WHEN THE PLASMALEMMAL IS DAMAGED. THIS FACTOR IS FIXED IN POINTS OF

DAMAGE TO THE VASCULAR WALL AND STARTS A CHAIN OF ENZYMATIC REACTIONS BETWEEN SPECIFIC PLASMA PROTEINS, LEADING TO THE FORMATION OF A DENSE CLOT. THE THERMOSTABLE COAGULATION ACTIVATOR INHIBITS THE MIGRATION OF FIBROCYTES INTO THE CLOT, HOWEVER, IT IS QUICKLY (DURING 3 – 5 MIN) CLEVAGED BY PLASMA LIPASE.

THERMOSTABLE COAGULATION ACTIVATOR CONCENTRATION DROPS AS SOON AS THE DAMAGE TO CELLS STOPS, AND THE CLOT IS ACTIVELY POPULATED BY FIBROCYTES OF THE VASCULAR WALL. FIBROCYTES LYSE CLOT PROTEINS AND REPLACE THEM WITH INTERCELLULAR SUBSTANCE TYPICAL FOR CONNECTIVE TISSUE (CLOT ORGANIZATION). FURTHER THE VESSEL LUMINAL CAN BE RECOVERED BECAUSE THE NEAREST ENDOMESOTHELIOCYTES OF THE VASCULAR WALL COVER THE ORGANIZED CLOT AND LYSE ITS INTERCELLULAR SUBSTANCE, FORMING CAVITIES.

PROTEINS PARTICIPATED IN THE PROCESS OF COAGULATION ARE ALSO SUBSTRATES FOR ENZYMES OF INTRAMESENTERIAL BODIES.

FORMED BLOOD ELEMENTS OF ALIENS ARE THREE TYPES OF SMALL CELLS.

1) MICROPHAGES – CELLS WITH A WIDE LIGHT CYTOPLASMA CONTAINING DUST-LIKE GRANULES AND A SMALL ROUND NUCLEUS. MICROPHAGES ARE ABLE TO PHAGOCYTOSIS OF BACTERIA, CELL DEBRIS, INERTE PARTICLES AND PENETRATION THROUGH VASCULAR WALLS INTO TISSUE. THEY ARE RESPONSIBLE FOR FIGHTING INFECTIOUS AGENTS AND IMPLEMENTING INFLAMMATORY REACTIONS. DIFFERENT TO THE NEUTROPHILS OF TERRESTRIAL VERTEBRATES, PHAGOCYTOSIS OF FOREIGN BODIES BY ALIEN MICROPHAGES IS NOT ASSOCIATED WITH THE GENERATION OF REACTIVE OXYGEN FORMS, BUT FREQUENTLY ENDS WITH THEIR MIGRATION INTO THE LUMINUM OF THE INTESTINE.

2) BASOPHILIIC CELLS – THE LARGEST FORMED BLOOD ELEMENTS OF ALIENS – ARE ROUND-SHAPED CELLS WITH BASOPHILIIC CYTOPLASMA CONTAINING LARGE OXYPHILIC GRANULES AND A SMALL ROUND NUCLEAR. BASOPHILE CELLS SYNTHESIZE ALL THE BASIC PLASMA WHITES, INCLUDING ALIENOCYANIN, AND ARE THE MAIN COPPER DEPO IN ALIEN'S BODY.

3) CELLS OF THE FAT BODY (SEE THE CONNECTIVE TISSUE OF ALIENS).

MICROPHAGES AND BASOPHILE CELLS ARE ABLE TO DEPOSIT IN SPECIAL AREAS OF THE CONNECTIVE TISSUE – DENSE BODIES – LOCATED IN THE THICKNESS OF THE COMMON MESENTERY. THE NUMBER AND SIZE OF DENSE BODIES ARE EXTREMELY VARIABLE. ACTIVE PEPTIDES OF INTRAMESENTERIAL BODIES AND BACTERIAL LIPOPOLYSACCHARIDES ARE STIMULATORS FOR THE EXIT OF THESE FORMED ELEMENTS FROM THE DENSE BODIES INTO THE VASCULAR BODY.

ALL FORMED BLOOD ELEMENTS OF ALIENS ARE ABLE FOR MITOTIC DIVISION IN THE LUMINAL OF THE VESSELS. MITOSIS STIMULATORS ARE BACTERIAL POLYSACCHARIDES AS WELL AS FACTORS PRODUCED BY FORMED ELEMENTS THEMSELVES.

PHYSIOLOGY OF CIRCULATION.

THE BLOOD FLOW IN ALIEN'S BODY IS PROVIDED BY THE «HEART» – TWO CUMULATIONS OF CIRCULATION-LOCATED SMOOTH MUSCLE CELLS IN THE WALLS OF THE ARTERIAL AND VENOUS CONNECTING DUCTS, THEY PULSE IN ANTI-PHASE. THE HEART OF ALIENS IS PULSING AUTOMATIC, HOWEVER, THE RITHM FREQUENCY IS REGULATED BY THE VISCERAL BRAIN THROUGH THE FIBERS OF THE COMMON VISCERAL NERVE, AND THE HUMORAL FACTORS OF INTRAMESENTERIAL BODIES. HIGH CONCENTRATIONS DECREASE THIS FREQUENCY AND LOW CONCENTRATIONS INCREASE IT.

VENOUS (DEXOXYGENIC) BLOOD, WHICH IS CURRENTLY FROM THE VEINS OF THE VENOUS SEGMENT OF THE PARACHORDAL RING, REACHING THE MOUTH OF THE VENOUS CONNECTING DUCT, IS PARTIALLY PUSHED BY THE PULSATIONS OF THE «HEART» INTO THE ANTENNAL SEGMENT OF THE BRANCHIAL RING, AND PARTIALLY FLOWS THROUGH THE THIRD SEGMENT TO THE ARTERIAL SEGMENT OF THE PACHORDAL RING, WHERE IT IS MIXED WITH AN ARTERIAL BLOOD. IN THE ANTENNAL SEGMENT OF THE BRANCHIAL RING, VENOUS BLOOD FROM THE VISCEROTHECA VEINS MIXES WITH VENOUS BLOOD FROM THE LOPOPHORUS VEINS. THEN THIS VENOUS BLOOD PASSES TO THE BRANCHIAL SEGMENT AND FURTHER FROM THE PULMONARY ARTERIES TO THE LUNGS. OXYGENATED BLOOD RETURNS TO THE BRANCHIAL SEGMENT THROUGH THE PULMONARY VEINS. CEREBRAL AND LOPOPHORUS ARTERIES BEGIN HERE. THEY SUPPLY OXYGENATED BLOOD TO THE UPPER LIMB, SENSOR ORGANS AND THE BRAIN. THE OTHER OXYGENATED BLOOD PARTIALLY GOES TO THE ARTERIAL SEGMENT OF THE PARACHORDAL RING, WHERE THE VISCEROTHECA ARTERIES BEGIN. THE OTHER PART GOES THROUGH THE THIRD SEGMENT OF THE BRANCHIAL RING TO THE ANTENNA SEGMENT, WHERE IT MERGES WITH THE DEOXYGENATED BLOOD OF THE VISCEROTHECA VEINS.

IT IS CLAIR FROM THE FOREGOING DESCRIPTION THAT MIXED BLOOD CIRCULATES IN THE ALIEN'S CIRCULATION SYSTE EXCEPT FOR PART OF THE BRANCHIAL RING. THE DEGREE OF BLOOD MIXING IS REGULATED BY THE REDUCTION OF SMOOTH MUSCLES INSIDE IN THE WALLS OF THE THIRD SEGMENTS OF THE PARACHORDAL AND BRANCHIAL RINGS. UNDER THE INFLUENCE OF NERVE AND HUMORAL IMPULSES, THE MIXING IN THE

PARACHORDAL AND BRANCHIAL ENDS CAN CHANGE FROM 0 TO 100% SEPARATELY, WHICH PROVIDES HIGH ADAPTABILITY.

PHYSIOLOGY OF NUTRITION AND DIGESTION.

MOST FORMS OF ALIEN ARE ABLE TO FEED ON THE ONLY FOOD THAT IS THE BIOMASS PRODUCED BY THE EATER. THIS PRODUCT HAS A DENCE CONSISTENCY AND OPTIMAL RATIO OF KEY NUTRIENTS, MAKING A FOOD PRE-TREATMENT SYSTEM NOT NEEDED. IN SUCH FORMS, THE GRAPTING AND EATING ORGAN PERFORMS THE FUNCTION OF ADDITIONAL LIMB. HOWEVER, THE HIGHEST ARE CAPABLE OF CANNIBALISM AND EATING ORGANIC RESIDUES. FEATURES OF ANATOMY AND PHYSIOLOGY OF THEIR DIGESTIVE SYSTEM ARE DESCRIBED…

… THE STOMACH OF MOST FORMS OF ALIEN (EXCEPT SOME SOLADES) PERFORMS A SINGLE FUNCTION – DEPOSITION OF FOOD RESERVES. THE STOMACH OF A SMALL WORKER IS ABLE TO HOLD UP TO 20 LITERS OF BIOMASS. UNDER THE ACTION OF NERVE STIMULAS FROM THE SIDE OF THE COMMON VISCERAL NERVE, THE LOWER SPHINCTER OF THE GASTRIC IS RELAXED PERIODICALLY, AND THE GASTROINTESTINAL CONTENT FALLS INTO THE SMALL INTESTINE.

THE MAIN PROCESSES OF DIGESTION AND ABSORPTION OCCUR IN THE SMALL INTESTINE OF ALIENS. MYOEPITHELIUM OF THE SMALL INTESTINE ACTIVELY SECRETES A MIXTURE OF HYDROLASE WITH PH OPTIMUM IN THE NEUTRAL REGION AND SUCKS THE RESULTING MONOMERS BY ENDOCYTOSIS OF THE INTESTINAL CONTENTS BY THE APICAL SURFACE OF THE EPITHELIAL COMPONENT AND THE EXOCYTOSIS FORMED VESICLES IN THE REGION OF THE CAPILLARIES UNDERGOING THE MUSCULAR COMPONENT OF MYOEPITHELIUM.

THE CONTENT OF THE VESICLES WITHIN THE CELL IS SEGREGATED; IN PARTICULAR, HIGH MOLECULAR SUBSTANCES – HYDROLASES AND NON-DESCIPLED POLYMERS – ARE SEPARATED FROM ENODCYTOSIC VESICLES INTO SPECIAL OBLONG-SHAPED BUBBLES WHICH ARE EXOCYTOSED INTO THE LUMINAL OF THE INTESTINE. THIS MECHANISM PREVENTS HYDROLASEMIA AND THE ABSORPTION OF VIRAL AGENTS, EXCEPT THE FACTOR OF ALLOCID-DEPENDENT TRANSDUCT. THE PROCESSES OF DIGESTION AND ABSORPTION ARE ACCELERATED BY COMPLEX PERISTALTIC MOVEMENTS OF THE MUSCLE COMPONENT OF THE MYOERYTHELIUM PROMOTING THE INTESTINAL CONTENT TOWARDS THE COLON. IN THE AREA OF THE TRANSITION OF THE SMALL INTESTINE INTO THE LARGE INTESTINE IS LOCATED A SPHINCTER, OPENING WHEN THE CONCENTRAUTION OF GLUCOSE IN THE INTESTINAL CONTENT REDUCES BELOW A SOME THRESHOLD LEVEL. GLANDS OF THE INTESTINAL WALL IN MOST FORMS OF ALIENS ARE IN THE STATE OF SUBATROPHY AND BEGIN

ACTIVE SECRETION ONLY WHEN THE INTESTINAL WALL STRETCHES DUE TO INCREASE IN THE VOLUME OF THE CONTENT ABOVE THE THRESHOLD LEVEL.

THE LARGE INTESTINE PERFORMS THE FUNCTION OF ABSORBING WATER, ELECTROLYTES, ESSENTIAL IONS, EXECRETION OF PHOSPHATE IONS AND LOADED MICROPHAGES. AS A RESULT OF THESE PROCESSES, FECES ARE FORMED – DRY SUBSTANCE (90% DRY RESIDUE) OF A GRAY-GREEN COLOR, CONSISTING OF NON-TREATED BIOMASS MICROORGANISMS (70%), DYSTROPHICALLY CHANGED MICROPHAGES LOADED WITH PHAGOCYTES (29%) AND POTASSIUM PHOSPHATE (1%) .

MOEPITHELIUS COLON HAS THE FOLLOWING BIOCHEMICAL FEATURES THAT ENSURE ITS BASIC FUNCTIONS:

1) ON THE APICAL MEMBRANE OF MYOEPITHELOCYTES ARE LOCALIZED BY THE CHEMOCH-SENSITIVE CHANNELS FOR WATER, ESSENTIAL CATIONS, POTASSIUM AND PHOSPHATE-ION, SENSITIVE TO INTRACITOPLASMIC CONCENTRATION OF THESE AGENTS, AS WELL AS A NUMBER OF HORMONES AND MEDIATORS OF THE GENERAL VISCERAL NERVE;

2) PROTON-DEPENDENT ANTIPORT SYSTEMS FOR THE SPECIFIED AGENTS ARE LOCALIZED IN RETICULAS BODIES MEMBRANES IN THE AREA OF THE MESENTERY CAPILLARIES;

3) MYOEPITHELIOCYTES ARE ABLE TO LET TO PASS THE MICROPHAGES LOADED WITH PHAGOCYTED MATERIAL FROM THE BLOOD INTO THE INTESTINAL LUMEN.

IT SHOULD BE NOTED THAT THE ENTIRE SMALL DIGESTIVE TRACT OF ALIENS, EXCEPT FOR SOME SPECIALIZED FORMS, DOES NOT HAVE A RESIDENT MICROFLORA.

THE ACT OF DEFECATION IN ALIENS IS INVOLVED AND IS PERFORMED BY SPASTIC REDUCTION OF THE TERMINAL SECTION OF THE SMALL INTESTINE, LAUNCHED BY INCREASING INTRACELLULAR PRESSURE AND COORDINATED WITH RELAXATION OF THE ANAL SPHINCTER. WHEN THE ANAL SPHINCTER IS RELAXED, THE FECES MIXES WITH THE MUCOUS SECRET GREEN GLAND.

GENERAL PHYSIOLOGY OF THE CENTRAL NERVOUS SYSTEM OF ALIENS.

AS IN TERRESTRIAL VERTEBRATES, INFORMATION PROCESSING AND MEMORIZATION PROCESSES IN THE CENTRAL NERVOUS SYSTEM (CNS) OF ALIEN ARE OCCURRED IN NEURAL NETWORKS. HOWEVER, IN CONNECTION WITH THE ELECTRIC MECHANISM OF INTERCELLULAR EXCITATION TRANSFER, THESE PROCESSES HAVE CERTAIN FEATURES.

IT IS KNOWN THAT THE SUBSTRATE OF THE CONDUCTION OF THE ACTION POTENTIAL IN ALIEN NEURONS ARE SPECIALIZED SECTIONS OF THE

PLASMALEMMA CONDUCTION BAND, WHERE POTENTIAL-DEPENDENT POTASSIUM CHANNELS ARE LOCALIZED. EACH NEURON PROCESS HAS FROM ONE TO FIVE CONDUCTION BANDS. IN THE AREA OF THE BODY OF A NEURON THE BANDS OF ITS PROCESSES INTERCEPSE, LEADING TO A COMPLEX INTERFERENCE OF ACTION POTENTIALS; A SIMILAR SITUATION OCCURS AT THE DISTAL ENDS OF NEURON PROCESSES. NOT ONLY SINGLE ACTION POTENTIALS INTERFERE, BUT ALSO PACKS OF DISCHARGE. THE LAWS OF INTERFERENCE CORRESPOND TO THE LAWS OF INTERFERENCE OF PERIODIC OSCILLATIONS. IF AS A RESULT OF THE INTERFERENCE OF ACTION POTENTIALS, THE MEMBRANE POTENTIAL OSCILLATION OCCURTS THAT DOES NOT EXCEED THE CRITICAL LEVEL OF DERPOLARIZATION, THE RESULT OF THE SUMMATION DEPENDS ON THE TOPIC OF ITS PLACE. IN THE AREA OF THE DISTAL ENDS OF THE PROCESSES, THESE VIBRATIONS ARE NEEDED OUT OF A TRACE, WHICH IS THE MAIN MECHANISM OF INDERATION IN THE CNS OF THE ALIENS. IN THE AREA OF THE NEURON BODY THE RESULT MAY BE THE SAME, HOWEVER, A SERIES OF NEURON POPULATIONS HAVE SPECIFIC MECHANISMS OF AMPLIFICATION OF SUBTHRESHOLD OSCILLATIONS OF THE MEMBRANE POTENTIAL:

1) IN THE NEURONS OF THE MAGNETOSENSOR NUCLEI OF THE OLIENCEPHALON BRAIN, POTENTIAL-SENSITIVE POTASSIUM CHANNELS ARE CAPABLE OF ADDITIONAL OPENING UNDER THE ACTION OF SHARP PERTURBATIONS OF THE MAGNETIC FIELD. THIS IS THE BASIS OF THE NEURONAL MECHANISM OF RECONSTRUCTION AND RECOGNITION OF MAGNETOSENSOR IMAGES. SOME NEURONS OF THE FINAL BRAIN HAVE A SIMILAR PROPERTY, BUT IT IS MUCH WEAKER.

2) IN THE NEURONS OF THE CENTRAL GRAY NUCLEUS OF THE VISCERAL BRAIN, HUMORAL CHEMOSENSORY STIMULES LEAD TO THE OPENING OF CHEMOSENSORY CALCIUM CHANNELS AND A DECREASE IN THE MEMBRANE POTENTIAL TO THE CRITICAL LEVEL OF DEPOLARIZATION.

3) IN THE NEURONS OF SOME AREAS OF THE OUTER LAYER OF THE GRAY MATTER OF THE SOMATIC BRAIN, THERMOGRANIN HAS A SIMILAR PROPERTIES. THERMO-GRANINE-DEPENDENT ACTIVATION OF THE DISCHARGE OF THESE NEURONS LEADS TO TREATMENT OF THE BODY INTEGUMENT MUSCLES AND INCREASED HEAT PRODUCTION.

4) NEUROSECRETORY STIMULES HAVE A SIMILAR IMPACT ON A SERIES OF NEURONS OF THE FINAL, OLIENCEPHALON AND SENSORY BRAIN. NEUROSECRETORY STIMULES CAN ALSO CAUSE A PHASE SHIFT OF THE ACTION POTENTIAL, WHICH CAN BE BOTH EXCITATORY AND INHIBITORY MECHANISMS. THESE EFFECTS ARE THE MAIN REALIZATION OF THE EMOTIONAL STATE OF ALIENS.

SYNAPSE BETWEEN ALIEN NEURONS ARE DIVIDED INTO TWO LARGE GROUPS:

1. SYNAPSES LOCALIZED IN PLACES OF POSSIBLE SUMMATION OF ACTION POTENTIALS ON THE SURFACE OF CELL BODIES AND DISTAL ENDS OF PROCESSES (ACTION POTENTIALS COMING THROUGH THESE SYNAPSE ARE SUMMER WITH OTHER ACTION POTENTIALS);

2. SYNAPSES FORMED ON THE SURFACE OF THE STRIPS OF PROCESSES – ACTION POTENTIALS COMING THROUGH THESE SYNAPSE, DEPENDING ON THE COMING CIRCUMSTANCES, CAN CAUSE THE FOLLOWING EFFECTS:

1) BLOCK OF CARRYING OUT THIS STRIP IN THE DISTAL DIRECTION FROM THE SYNAPSE;

2) RETROGRADE DISTRIBUTION OF THE INCOMING ACTION POTENTIALS (TO THE CELL BODY) OR ITS ANTEOGRADE DISTRIBUTION (TO THE DISTAL PART OF THE PROCESS);

3) AMPLIFICATION OF SUBTHRESHOLD OSCILLATIONS OF THE MEMBRANE POTENTIAL OF THE BAND CAUSED BY THE OPENING OF CHEMOSENSITIVE CALCIUM CHANNELS UNDER THE ACTION OF NEUROSECRETORY OR HUMORAL STIMULES TO THE CRITICAL LEVEL OF DEPOLARIZATION.

4) THUS, THE PROCESSES OF EXCITATION AND INHIBITION IN THE NERVOUS SYSTEM OF ALIENS CAN BE «PRE-SUMMING « (OCCURRING WITHIN A SINGLE CONTROL STRIP) AND «INTRA-SUMMING» (OCCURRING IN THE POINT OF INTERCROSSING SEVERAL CONDITION STRIPS).

THE BASIS OF THE MEMORIZATION PROCESSES IN ALIENS IS:

1) CHANGING THE GEOMETRY OF THE EXISTING BANDS AND SUMMATION AREAS;

2) FORMATION OF NEW SYNAPTIC CONTACTS;

3) SELECTIVE CHANGES IN THE LEVEL OF POTENTIAL AND CHEMOSENSITIVITY OF ION CHANNELS WITH A CONSTANT GEOMETRY OF SYNAPSE AND CONDUCT BANDS (THE BASIS OF SHORT-TERM MEMORY).

(THE FIRST TWO MECHANISMS ARE THE BASIS OF LONG-TERM MEMORY).

THE FORMATION OF NEW SYNAPTIC CONTACTS OCCURRS AS A RESULT OF THE INTERACTION OF TWO PROCESSES:

1) THE PHENOMENON OF SPROUTING – CONTINUOUS FORMATION OF CYTOPLASMATIC OUTGOINGS CARRYING CONNEXON-LIKE STRUCTURES IN THE AREA OF SUMMATION AREAS;

2) THE PHENOMENON OF THE SELECTIVE SEARCH OF THE RECEPTIVE STRUCTURE, CONSISTING IN THAT WHEN THE CYTOPLASMATIC PROCESSES OF TWO CELLS MEET, A TEMPORARY CONTACT IS FORMED. IF ACTION POTENCIAL COMING BY THE PROCESSES SATISFIATE WITH SOME NOT COMPLETELY CLEAR

CONDITIONS, DIFFERENT FOR EACH NEURAL NETWORK, AFTER A FEW DAYS A TYPICAL ELECTRIC SYNAPSE IS FORMED IN THE PLACE OF CONTACT; IN THE OPPOSITE CASE, THE PROCESSES ARE REDUCED.

FROM THE FOREGOING, IT IS CLEAR THAT THE APPEARANCE OF NEW SYNAPSE IS USUALLY PREDICTED BY THE APPEARANCE OF NEW SUMMATION AREAS; ONLY TYPE I SYNAPSES ARE NEWLY FORMED.

CHANGING THE GEOMETRY OF CONDUCT BANDS IS CONNECTED TO LASTLY WITH CHANGES IN THE STRUCTURE OF THE SUBMEMBRANE ELEMENTS OF THE CYTOSKELETON. IT MAY RESULT IN: 1) CHANGING THE RESULTS OF THE SUMMATION OF THE ACTION POTENTIAL; 2) FORMATION OF NEW SECTIONS OF SUMMATION.

CHANGING THE GEOMETRY OF SUMMATION BANDS IS THE MAIN MECHANISM OF LONG-TERM MEMORY. UNDER THE ACTION OF NEUROSECRETORY AND HUMORAL-PHEROMONE STIMULES, EXTREMELY RAPID CHANGES IN THE GEOMETRY OF CONDUCT BANDS CAN OCCUR, THAT IS THE BASIS OF THE PHENOMENON OF EIDETIC IMPRINTING (SEE BELOW).

<u>THE CONNECTION OF THE NERVOUS SYSTEM WITH THE EFFECTOR ORGANS</u> IN ALIENS IS IMPLEMENTED IN TWO WAYS – ELECTRIC AND CHEMICAL. ELECTRICAL COMMUNICATION CONSISTS IN FORMATION PERIPHERAL NERVES BY LONG PROCESSES OF NEURONS. TYPICAL IS ELECTRICAL SYNAPSE WITH THE MUSCLE COMPONENT OF THE MYOEPITHELY. THESE PROCESSES HAVE ONE CONDUCT BAND AND ARE ABLE TO FORM ELECTRICAL SYNAPSE SIMULTANEOUSLY WITH SEVERAL MYOEPITHELIOCYTES (MOTOR UNIT). THE MUSCLES OF BODY COVERS, LIMBS, LUNGS, PSEUDOSYRINX, THE MUSCLE COMPONENT OF THE GASTROINTESTINAL TRACT, DRAINAGE DUCTS, GREEN AND EXOCRINETIC GLANDS ARE INNERVATED IN A SIMILAR WAY.

CHEMICAL COMMUNICATION IS MOSTLY LIMITED TO THE COMMON VISCERAL NERVE SYSTEM. PART OF THE COMMON VISCERAL NERVE FIBER DOES NOT FORM ELECTRICAL SYNAPSE, BUT SECRETE A SPECIFIC OLIGOPEPTIDE INTO THE INTERCELLULAR FLUID OF EFFECTIVE TISSUES AND INTO VESSEL LUMINIES. IT IS A HUMORAL FACTOR AFFECTING EFFECTIVE CELLS BY THE ACTIVATION OF CHEMOSENSITIVE CALCIUM CHANNELS. THIS WAY THE EPITHELIAL COMPONENT OF TYPE I-IV MYOEPITHELIA, PARENCHYMA OF THE GREEN GLAND AND EXOCRIN GLANDS, SMOOTH MUSCLE OF THE HEART AND VESSELS ARE «INNERVED». EXCEPTIONS ARE THE INTERSYRINGIAL GLANDS AND THE MUSCLES OF THE SPHINCTERS OF INTERMESENTERIAL BODIES. (THE INTERSYRINGIAL GLANDS ARE ELECTRICLY INNERVATED BY THE NERVE FIBERS OF SPIATED PLEXUS ISSUING

FROM THE SOMATIC BRAIN. ONLY SPHINCTERS OF INTERMESENTERIAL BODIES ARE ELECTRICALLY INNERVED SMOOTH MUSCLES).

<u>THE ROLE OF CHEMICAL STIMULUSES</u> IN THE CENTRAL NERVOUS SYSTEM OF ALIENS IS NOT A ROLE OF EXCITATION MEDIATORS, BUT ITS MODULATORS. HERE THE MAIN ROLE IS NOT BELONG TO EXCITATION MEDIATORS BUT TO MODULATORS. THE GENERAL MOLECULAR MECHANISM IS THE ACTIVATION OF CHEMOSENSITIVE CALCIUM CHANNELS IN VARIOUS NEURONS, CAUSING THE FOLLOWING EFFECTS:

1) CLOSURE OF CALCIUM-DEPENDENT POTASSIUM CHANNELS AND GENERATION OF ACTION POTENTIAL (THIS IS CHARACTERISTIC FOR THE CHEMOSENSORY STRUCTURES OF THE VISCERAL BRAIN AND SOME NEUROSECRETORY STRUCTURES);

2) GRADUAL DEPOLARIZATION OF SUMMATION SECTIONS, INCREASING THE SUBTHRESHOLD OSCILLATIONS OF THEIR MEMBRANE POTENTIAL TO THE CRITICAL LEVEL OF DEPOLARIZATION;

3) THE PHENOMENON OF ACTION POTENTIAL PHASE SHIFT, CONSISTING IN THAT ON THE BACKGROUND OF THE DECREASING PHASE OF THE ACTION POTENTIAL, AN ADDITIONAL WAVE OF DEPOLARIZATION APPEARS, DELAYING THE RECOVERY OF THE MEMBRANE POTENTIAL.

4) RESTRUCTURING OF THE GEOMETRY OF CONDUCT BANDS UNDER THE INFLUENCE OF A SINGLE VOLVO OF DISCHARGE SYNCHRONOUS WITH THE OPENING OF CALCIUM CHANNELS. THIS IS THE BASIS OF THE EIDETIC IMPRINTING EFFECT;

5) EXOCYTOSIS OF VESICLES CONTAINING NEUROSECRETORY FACTORS.

CHEMICAL STIMULUSES AFFECTING THE CENTRAL NERVOUS SYSTEM OF ALIENS CAN BE DIVIDED INTO THREE GROUPS:

1. HUMORAL STIMULUSES – GASES, HORMONES AND BLOOD METABOLITES, ACTING MOSTLY ON THE GRAY NUCLEUS NEURONS OF THE VISCERAL BRAIN, CAUSING HOMEOSTATIC REACTIONS, AS WELL AS CHANGES IN THE SECRETION OF NEUROSECRETORY FACTORS.

2. PHEROMONO-HUMORAL STIMULUSES. NEUROEPITHELIOCYTES OF BEADED CUPULS HAVE THE ABILITY TO CAPTURE PHEROMONES SORBED ON MUSCUS, AND TRANSPORT THEM BY OLFACTORY FIBERS AND SECRETE INTO THE FLUID FILLING THE BRAIN CAPSULE. SUCH TRANSPORT IS MAJORLY EXPOSED TO THE PHEROMONES OF THE STATUS GLANDS, THE HYPOPHARYNGEAL BODY, THE COXAL GLANDS, AS WELL AS SOME PHEROMONES OF THE INTRASYRINGEAL GLANDS, RECOVERED IN THE HIGHEST CONCENTRATIONS. PHEROMONO-HUMORALDIC STIMULES CAN CAUSE THE FOLLOWING EFFECTS: 1) VARIOUS STEREOTYPE MOTOR REACTIONS – FEEDING, CATALEPSY, ATTACKS, ETC.; 2) THE

PHENOMENON OF EIDETIC IMPRINTING – MEMORIZING OR FORGETTING INFORMATION COMBINED WITH A PARTICULAR SMELL CODE FROM ONE PRESENTATION. THE SMELL CODE SHOULD ALWAYS INCLUDE STATUS PHEROMONES OF A UPPER CASTE INDIVIDUAL THAT IS HIGHER STANDING ON THE HIERARCHICAL LADDER; 3) CHANGES IN THE SECRETION OF NEUROSECRETORY FACTORS.

A SERIES OF ALIEN NEURONS IN RESPONSE TO THE IMPACT OF VARIOUS COMBINATIONS OF PHEROMONES OF SUPERIOR INDIVIDUALS, ARE ABLE TO SYNTHESIZE AND RELEASE PEPTIDE COMPOUNDS, CAUSING THE FOLLOWING EFFECTS:

1) SUBJECTIVE CHANGE OF EMOTIONAL STATE;

2) CHANGES IN SENSITIVITY, SPEED AND NATURE OF INFORMATION PROCESSING IN SENSORY SYSTEMS OF THE BRAIN;

3) INDUCTION OF THE STATE OF PRESETTING TO MOVEMENTS OF A CERTAIN TYPE;

4) CHANGING THE FUNCTIONS OF THE INTERNAL ORGANS;

5) FACILITATION OR DELAY OF PROCESSES OF MEMORIZATION AND FORGETTING.

THERE ARE THREE TYPES OF SUCH NEURONS.

TYPE «A» CELLS ARE LOCALIZED IN THE OUTER LAYER OF THE BRAIN SUBSTANCE OF THE VISCERAL BRAIN. SHORT PROCESSES OF THESE CELLS CONTACT WITH CHEMOSENSORY NEURONS OF THE CENTRAL GRAY NUCLEUS OF THE VISCERAL BRAIN; LONG PROCESSES DISTRIBUTED FFUZLY IN THE WHITE AND GRAY SUBSTANCE OF THE TERMINAL, SENSORY AND OLIENCEPHALON BRAIN. ON THE BODIES OF TYPE «A» CELLS THERE ARE SYNAPSE WITH PROCESSES OF TYPE «B» CELLS. TYPE «A» CELLS SECRET OLIGOPEPTIDE «A» HAVING THE FOLLOWING EFFECTS (WHEN INTRODUCED INTO THE BRAIN CAPSULE):

1) INDUCTION OF A SUBJECTIVE STATE OF BLISSED PEACE/SATISFACTION/UNWILLING TO MOVE/FEEL/THINK;

2) INDUCTION OF HALLUCINATIONS OF A PLEASANT CHARACTER, BY THE CONTENT ASSOCIATED WITH THE NUTRITION PROCESS (IN EXTREMELY HIGH DOSES);

3) LIMITATION OF THE RECEIVING ABILITY OF DISTANT RECEPTORS, STRENGTHENING OF THE RECEIVING ABILITY OF CONTACT AND INTER RECEPTORS;

4) STRENGTHENING SECRETION AND PERISTALSIS OF THE GASTROINTESTINAL TRACT, RESPIRATORY IMPAIRMENT, INCREASING THE DURATION OF THE DIFFUSION PAUSE, REDUCED HEART RATE, REDUCING THE

SMOOTH MUSCLES OF THE THIRD SEGMENT OF THE BRANCHIAL RING (LEADS TO THE SUPPLY OF VISCERAL ORGANS AND LOWER LIMBS BY OXYGENATED BLOOD), BRAKING THE SECRETION OF THE SUBPHARINGEAL BODIES;

5) CREATION OF A STATE OF PRE-ADJUSTMENT TO AN ELEMENTARY SENSORIMOTOR REACTION TO A STRONG CONTACT IRRITIVE (THE REACTION APPEARS IN A STRONG JUMP UP AND TO A SIDE) ON THE BACKGROUND OF A DECREASE IN THE GENERAL MUSCLE TONE;

6) INHIBITION OF PROCESSES OF FORMATION OF LONG-TERM MEMORY, PARTICULARLY FOR COMPLEX STIMULES OF SPEECH CHARACTER.

THE CELLS OF THE TYPE «B» ARE LOCALIZED ON THE LOWER SURFACE OF THE SENSORY BRAIN (SHORT PROCESSES CONTACT WITH THE NEURONS OF THE OUTER LAYER OF THE GRAY SUBSTANCE), IN THE OUTER LAYER OF THE GRAY SUBSTANCE OF THE SOMATIC BRAIN (SHORT PROCESSES CONTACT WITH THE NEURONS OF THE STRUCTURES RESPONSIBLE FOR THE PROPRIOSEPSION, THE PERCEPTION OF SOUND AND VIBRATION STIMULUS) IN WHITE MATTER OF THE BRAIN. THE LONG PROCESS OF THE «B» CELLS CONTACT WITH NEURONS OF THE TERMINAL BRAIN, MAGNETOSENSOR NUCLEUSES, NEUROEPITHELIUM OF OCELLIS AND BEADES CULULS (PASSING BY ANTENNAL AND OCELLI NERVES), WITH NEURONS OF THE OUTER LAYER OF THE GRAY SUBSTANCE AND THE CENTRAL LAYER OF THE SOMATIC BRAIN, WITH CHEMOSENSORY NEURONS, WITH CELLS OF TYPE «A» AND WITH NEURONS OF THE OUTER LAYER OF THE GRAY SUBSTANCE OF THE VISCERAL BRAIN.

THESE PROCESS OF TYPE «B» CELLS, IN ADDITION TO CONTACTING CELLS OF THE CELLS OF TYPE «A», SECRETE OLIGOPEPTIDE «B». WHEN IT IS INGECTED TO THE BRAIN CAPSULE LIQUID, THE FOLLOWING EFFECTS ARE:

1) INDUCTION OF THE SUBJECTIVE EMOTIONAL STATE OF FEAR / DISSATISFACTION / ASPIRATION FOR AGGRESSION / THIRST FOR DOMINANCE;

2) INDUCTION OF HALLUCINATIONS OF A TERRIBLE CHARACTER, BY THE CONTENT ASSOCIATED WITH REACTIONS OF AGGRESSION OR FLIGHT (IN EXTREMELY HIGH CONCENTRATIONS);

3) SIGNIFICANT INCREASE IN SENSITIVITY OF ALL DISTANT RECEPTORS, DECREASE IN SENSITIVITY OF PROPRIORECEPTORS, INTERORECEPTERS, CHEMOSENSORY NEURONS, INCREASED THRESHOLD OF PAIN REACTION;

4) BRAKING PERISTALSIS AND SECRETION OF THE GASTROINTESTINAL TRACT, LOWERING THE PRESSURE THRESHOLD FOR THE ACT OF DEFECATION, THE INCREASE IN RESPIRATION, INCREASING THE HEART RATE, CONTRACTION OF THE MUSCLES OF THE THIRD SEGMENT OF THE BRANCHIAL RING (IN COMBINATION WITH OTHER REACTIONS LEADS TO AN INCREASE IN THE AMOUNT OF BLOOD OXYGENATED PER UNIT TIME), INCREASED SECRETION OF THE

HYPOPHARYNGEAL BODY, STATUS GLANDS AND COXAL GLANDS, INCREASED CONCENTRATION OF THE SECRETION OF INTRASYRINGEAL GLANDS;

5) INDUCTION OF THE STATE OF PRESETTING TO THE STANDARD COMPLEXES OF COMBAT MOVEMENTS (IN SOLDIERS), AS WELL AS RUNNING, ON THE BACKGROUND OF INCREASING THE GENERAL MUSCLE TONE;

6) STRENGTHENING THE PROCESSES OF MEMORIZING SIMPLE STIMULES, AS WELL AS SPEECH STIMULUS OF THE COMPLEX CHARACTER OF THE SUPERIOR CAST (EIDETIC IMPRINTING OF ORDERS).

TYPE «C» CELLS ARE LOCALIZED IN THE WHITE MATTER OF THE TERMINAL BRAIN, THE MAGNETOSENSOR NUCLEI OF THE OLIENCEPHALON BRAIN, THE OUTER LAYER OF THE GRAY MATTER OF THE SOMATIC BRAIN. THE PROCESSES OF THESE CELLS CONTACT WITH THE CELLS OF THE TERMINAL AND SENSORY BRAIN, THE OUTER LAYER OF THE GRAY MATTER OF THE OLIENCEPHALON BRAIN. TYPE «C» CELLS SECRETE OLIGOPEPTIDE «C», BEIN INGECTED TO THE LIQUIDE OF BRAIN CAPSULE , CAUSING THE FOLLOWING EFFECTS:

1) INDUCTION OF THE SUBJECTIVE EMOTIONAL STATE OF CALM AND INDIFFERENCE IN SOLDIERS, CURIOSITY AND DESIRE;

2) HALLUCINATIONS OF EMOTIONALLY NEUTRAL CHARACTER, IN WORKING AND SOLDIERS ASSOCIATED WITH PROFESSIONAL ACTIVITIES, AND IN UPPER CASTES IT IS ASSOCIATED WITH RELIGIOUS CEREMONIES;

3) REDUCED SENSITIVITY OF PROPRIORECEPTORS, INTERORECEPTERS, CHEMOSENSORY NEURONS, INCREASED THRESHOLD OF PAIN REACTION;

4) INDUCTION OF THE STATE OF PRESETTING TO THE STANDARD COMPLEXES OF LABOR MOVEMENTS IN LOW-INTELLIGENT CASTS, RELEASE OF ASSOCIATIVE ACTIVITY IN HIGHLY INTELLIGENT CASTS;

5) SHARP INCREASE IN THE PROCESSES OF MEMORIZING EMOTIONALLY NEUTRAL INFORMATION ON THE BACKGROUND OF DEACTUALIZATION OF EMOTIONAL STIMULES.

OLIGOPEPTIDE «C» IS FORMED NOT ONLY WHEN ACTIVATION OF TYPE «C» CELLS, BUT ALSO AS A RESULT OF CLEAVAGE OF OLIGOPEPTIDE «B» BY A SPECIFIC PEPTIDASE. THIS ENZYME IS CONTAINED IN THE ENDOTHELIUM OF THE BRAIN VESSELS AND IS SECRETED INTO THE FLUID OF THE BRAIN CAPSULE IF THE BLOOD PRESSURE IN THE PARACHORDAL RING IS WITHIN CERTAIN LIMITS. SO IN VIVO IN A NUMBER OF CERTAIN SITUATIONS, MOSTLY IN THE PRESENTATION OF ADVERSIVE STIMULES OF MODERATE STRENGTH, PHEROMONE AND MAGNETOSENSOR CONTACTS WITH A SUPERIOR INDIVIDUAL, INCREASING THE INTRACEREBRAL LEVEL OF OLIGOPEPTIDE «B» IS ACCOMPANIED WITH

INCREASING OF OLIGOPEPTIDE «B». INGECTION OF THE MIXTURE OF AGENTS IN THE LIQUID OF THE BRAIN CAPSULE CAUSES THE FOLLOWING EFFECTS:

1) INDUCTION OF THE SUBJECTIVE EMOTIONAL STATE OF COMPLETE SUBORDINATION TO UPPER CASTES/ AGGRESSION IN RELATION TO THE LOWER CASTES AND ESPECIALLY FORMS OF LIFE, NOT RELATED TO THE NATIVE NEST / THIRST FOR WORK (IN WORKING CASTS) / BATTLE (IN SOLDIERS) /THE MENTAL ACTIVITY (IN UPPER CASTES);

2) A SIGNIFICANT INCREASE IN THE SENSITIVITY OF ALL DISTANT RECEPTORS AND PROPRIORECEPTORS, A DECREASED SENSITIVITY OF INTERORECEPTERS, CHEMOSENSORY NEURONS, INCREASED THRESHOLD OF PAIN SENSITIVITY;

3) INDUCTION OF THE STATE OF PRESETTING TO THE STANDARD COMPLEXES OF LABOR/COMBAT MOVEMENTS IN LOW-INTELLIGENT CASTS, AND THE ASSOCIATIVE ACTIVITY IN UPPER CASTES;

4) SHARP STRENGTHENING OF THE PROCESSES OF MEMORIZING BOTH EMOTIONALLY SIGNIFICANT AND NEUTRAL INFORMATION.

<u>PHYSIOLOGY OF SENSE ORGANS AND SPEECH FUNCTIONS OF ALIENS.</u>

VISION. SIMPLE EYES OF INDIVIDUALS WITHOUT OBJECT VISION ARE HIGHLY SENSITIVE LIGHT LEVEL DETECTORS. THIS FUNCTION IS IMPORTANT FOR THE IMPLEMENTATION OF FLIGHT REACTIONS (A FEARED ALIEN STRIVES TO BE IN THE LOST LIGHT PLACE), AND ALSO FOR THE IDENTIFICATION OF BUILDINGS (IN COMPLEX WITH OLFACTORY STIMULES). IN ADDITION, THE «EYE – SUBMACULAR ORGANS» COMPLEX IS RESPONSIBLE FOR THE REGULATION OF METABOLISM CIRCAD RHYTHMS AND CHANGES IN WATER-SALT METABOLISM WHEN GOING TO OPEN SPACE.

COMPOUND EYES OF INDIVIDUALS WITH OBJECT VISION CONSIST OF SET (15000 – 16000 SIMPLE EYES). THE RESOLUTION OF THE COMPOUND EYE IS APPROXIMATELY 20', WHICH IS 20 TIMES LOWER THE RESOLUTION OF THE HUMAN EYE. THE LOCATION OF THE NEAREST POINT OF CLEAR VIEW IN DIFFERENT INDIVIDUALS IS DETERMINED BY THE CURVATURE OF THE COVERING LENS. ALIENS HAVE BINOCULAR VISION AND THE ABILITY TO JUDGE THE DISTANCE TO OBJECTS, HOWEVER, IT IS IMPERFECT DUE TO THE LACK OF SEPARATE EYE MOVEMENTS. THE VISUAL RECEPTION OF THE ALIENS SERVES AS A SUPPLEMENT TO MAGNETOSENSORS.

DISTANT CHEMORECEPTION (THE EQUIVALENT OF OLFACTORY RECEPTION IN MAMMALS) IN ALIENS IS CHARACTERIZED BY AN EXTREMELY HIGH AND NARROWLY SELECTIVE SENSITIVITY. THE SPECTRUM OF RECEPTABLE MOLECULES IS VERY LIMITED – THESE ARE THE SECRETS OF THE EXOCREEN

GLANDS, CARBON DIOXIDE, WATER, AMMONIA, PUTRESCINE, HYDROGEN SULFIDE, AND VARIOUS LOW-MOLECULAR SUBSTANCES. HOWEVER, ALIENS ARE ABLE TO RECOGNIZE OPTICAL ISOMERS OF MOLECULES, WHICH IS THE BASIS FOR RECOGNIZING THE CAST STATUS AND INTERPRETATION OF SPEECH PHEROMONE COMPONENTS.

SOME MOLECULES THAT CONTAIN THE SECRETS OF EXOCRINE GLANDS ACCUMULATE ON THE MEMBRANES OF SENSOR NEUROEPITHELIOCYTES OF ROSE CAPSULES, ARE ABSORBED IN THE COMPOSITION OF ENDOCYTOSIS VESICLES AND TRANSPORT BY THE FIBERS OF THE ANTENNALIA NERVE. THESE AGENTS HAVE A DIRECT EFFECT ON NEUROSECRETORY CELLS AND THE CELLS OF THE OUTER LAYER OF THE GRAY MATTER OF THE OLIENCEPHALON BRAIN, CAUSING CHANGES IN THE EMOTIONAL STATE (MAINLY BY THE TYPE OF JOINT INCREASE IN THE LEVEL OF OLIGOPEPTIDES «IN» AND «C») AND STEREOTYPICAL MOTOR REACTIONS (THE SMELL OF THE SECRET OF THE HYPOPHARYNGEAL GLAND OF INDIVIDUALS OF A SUPERIOR CASTE, AT A CERTAIN ITS STRENGTH, CAUSES IN INDIVIDUALS OF WORKING CASTES AN INVOLVED BURP REACTION.

Printed in Great Britain
by Amazon